Dr. Akhil Gupta

Solar Photovoltaics Engineering

A Power Quality Analysis Using Matlab Simulation Case Studies

Anchor Academic
Publishing

Gupta, Akhil: Solar Photovoltaics Engineering. A Power Quality Analysis Using Matlab Simulation Case Studies, Hamburg, Anchor Academic Publishing 2016

Buch-ISBN: 978-3-96067-082-7
PDF-eBook-ISBN: 978-3-96067-582-2
Druck/Herstellung: Anchor Academic Publishing, Hamburg, 2016

Bibliografische Information der Deutschen Nationalbibliothek:
Die Deutsche Nationalbibliothek verzeichnet diese Publikation in der Deutschen Nationalbibliografie; detaillierte bibliografische Daten sind im Internet über http://dnb.d-nb.de abrufbar.

Bibliographical Information of the German National Library:
The German National Library lists this publication in the German National Bibliography. Detailed bibliographic data can be found at: http://dnb.d-nb.de

All rights reserved. This publication may not be reproduced, stored in a retrieval system or transmitted, in any form or by any means, electronic, mechanical, photocopying, recording or otherwise, without the prior permission of the publishers.

Das Werk einschließlich aller seiner Teile ist urheberrechtlich geschützt. Jede Verwertung außerhalb der Grenzen des Urheberrechtsgesetzes ist ohne Zustimmung des Verlages unzulässig und strafbar. Dies gilt insbesondere für Vervielfältigungen, Übersetzungen, Mikroverfilmungen und die Einspeicherung und Bearbeitung in elektronischen Systemen.

Die Wiedergabe von Gebrauchsnamen, Handelsnamen, Warenbezeichnungen usw. in diesem Werk berechtigt auch ohne besondere Kennzeichnung nicht zu der Annahme, dass solche Namen im Sinne der Warenzeichen- und Markenschutz-Gesetzgebung als frei zu betrachten wären und daher von jedermann benutzt werden dürften.

Die Informationen in diesem Werk wurden mit Sorgfalt erarbeitet. Dennoch können Fehler nicht vollständig ausgeschlossen werden und die Diplomica Verlag GmbH, die Autoren oder Übersetzer übernehmen keine juristische Verantwortung oder irgendeine Haftung für evtl. verbliebene fehlerhafte Angaben und deren Folgen.

Alle Rechte vorbehalten

© Anchor Academic Publishing, Imprint der Diplomica Verlag GmbH
Hermannstal 119k, 22119 Hamburg
http://www.diplomica-verlag.de, Hamburg 2016
Printed in Germany

Acknowledgement

I am thankful to Dr. Saurabh Chanana, my Ph.d. supervisor at Department of Electrical Engineering, National Institute of Technology, Kurukshetra, Haryana, India, for his kind support which he rendered out of his busy schedule during the course of my earlier research work. His approach and ability to derive the solutions to any problem in this world has immensely helped me in increasing my ability during finding solutions to problems. It was only after my research work had been over, I was able to present this manuscript with a focused approach.

I am indebted to my parents, my wife Ms. Yachana Gupta and my children. Special thanks at extreme end to my wife for her valuable support time to time which made me free out of other social activities to give enough time in writing this book.

Finally,
A big Thanks are due to my other family members, my teachers, my students,
my colleagues, my friends and The God
Special Thanks to Diplomica-Verlag Publishing House

Dedicated to
MY FAMILY

Preface

This book comprises 10 chapters each of which covers the study on solar energy and its application in solar Photovoltaic grid connected systems. Power quality investigation has been carried highlighting the study of key power quality parameters. Author has made an endeavor to tailor the all chapters for an audience of university faculty members, students, researchers and practitioners in solar energy sector.

Chapter 1 mainly presents the world perspective and growth of electric power through conventional sources and its vis-à-vis comparative growth with various sources of renewable energy. It has been discussed that electric power has been one of the most critical component among others available required for infrastructure and crucial for economic growth and welfare of any nation. The growth of any developing nation mainly depends on various resources critically required in growth of infrastructure. This chapter also highlights the recent advancements recently taken place in renewable energy. The measures being adopted for building and supporting in the further renewable energy growth in developing and developed economies is also presented.

In Chapter 2, an introduction of a solar PV cell with its behavior with the help of its electrical circuits in ideal and practical conditions is presented. This description is carried out using its basic output current and output voltage equations. In addition, the various parameters as defined in a standard manufacturer data-sheet is discussed. The basic difference between a solar PV cell and a normal p-n junction diode is also described. The importance of MPP using V-I and P-V electrical characteristics is highlighted. In order to study any PV model, the dependence of a solar PV cell on variable solar radiation and ambient temperature is derived and discussed.

Chapter 3 has presented a discussion on advantages and limitations of crystalline and thin film technologies with their common features. This chapter has begun from by looking at the development of solar cells from the early stage. Then the commercial silicon wafer based solar cell technology is described. Special attention has been given to each thin film technology as its related material properties, fabrication techniques are discussed. The chapter has covered thin film technologies such as amorphous silicon, CdTe, CIGS, and thin film crystalline silicon solar cell technologies.

Chapter 4 has highlighted the needs of PV installation components other than PV panels. These components are jointly referred to as the Balance of System (BOS) and include the batteries, DC-DC converters, DC-AC converters for AC loads and grid connected systems. Since *P-V* and *V-I* curves of a solar panel are non-linear in nature, the need and importance of MPPT technique in extracting variable solar power at maximum points is described in PCS systems.

In chapter 5, the introduction of PQ and its concerns has been presented. The various PQ issues have been introduced and discussed according to IEEE and IEC standards. The effect of each of PQ issue has been described for a distribution system according to IEC standard. It is discussed that in electric power systems, hundreds of power generating stations and load centers are interconnected and operated. There are various PQ problems likewise, voltage sag, voltage swell, transients, voltage interruption, harmonics, noise and notching. Each PQ problem can be solved by the proper coordination among all power system components.

Chapter 6 has presented that Matlab is a software package, which can be used to perform analysis and solve mathematical and engineering problems. Introduction to various Matlab windows have been given and described. Simulink contains a library editor of tools from which we can build input/output devices and continuous and discrete time model simulations. Simulink has a comprehensive block library which can be used to simulate linear, non-linear or discrete systems. C codes can also be generated from Simulink models for embedded applications and rapid prototyping of control systems. This chapter also highlights the procedure for the operation of Simulink software with building blocks.

Chapter 7 has presented the modeling of solar PV single stage grid connected system at unity power factor. No transformer is used in the proposed system as it increases the level of harmonics in the overall system. The nature of real power generated by solar PV array through VSC has been shown and proved that whenever the power from grid is un-available, the real power requirement of the load is achieved by VSC. Data based MPPT is proposed through which behavior of actual DC link voltage is discussed.

In chapter 8, the simulation of the proposed system is performed under two different cases. In first case, the effect of changing power factor on active power, reactive power and THD values is observed. It is observed that the THD of grid current increases with increase in the phase angle of grid current VSC voltage. It affects the active power flow among VSC, load and utility grid. In the second case, the effect of changing frequency on active power, reactive power and THD values is noticed. It is observed that the THD of VSC current increases whereas, the THD of grid current remains constant.

Chapter 9 has presented the analysis of two-stage solar PV grid connected system which is evaluated at linear *RLC* load. In DC-DC boost converter, the IC-MPPT technique which is capable to operate even under changing environmental conditions is implemented. Real and reactive power exchange is exchanged among VSC, load and utility grid. Voltage and current waveforms are presented. In order to evaluate the level of power quality, the THD analysis is carried out using FFT. It has been found that although level of harmonics generation from VSC is high, the control system is designed that level of harmonics is reduced for grid injected current.

Chapter 10 has discussed the system performance at P&O based MPPT technique for solar PV two-stage grid connected system. It has been found that harmonic level is reduced for converter current and grid injected current. However, this MPPT is not able to track the reference MPPT voltage accurately. This validates that this MPPT is not able to operate under wide range of changing environmental conditions. The behavior of voltage and current levels of VSC, load connected and utility grid has also been discussed. Active and reactive power exchange among VSC, load connected and utility grid has also been highlighted.

Table of Contents

Acknowledgement 5

Preface 7

List of Abbreviations 14

List of Tables 16

List of Figures 17

CHAPTER I: ELECTRIC POWER - A WORLD PERSPECTIVE 21

    1.1. Introduction to Power Sector-A World Perspective 21

        1.1.1 Coal 22

        1.1.2 Natural gas 22

        1.1.3 Petroleum and Other Liquid Fuels 22

        1.1.4 Renewable Resources 22

    1.2 Status of Indian Power Sector 23

    1.3 Status of Global Renewable Energy 25

    1.4 Conclusion 27

CHAPTER II: AN INTRODUCTION TO SOLAR PHOTOVOLTAIC CELL & MODELING 28

    2.1 Introduction 28

    2.2 A Solar PV Cell-A p-n Junction Diode 28

    2.3 Difference Between a Solar PV Cell and a p-n Junction Diode 30

    2.4 Basic Parameters of Solar PV Cell 31

        2.4.1 Voltage-Current (V-I) Characteristics 31

        2.4.2 Short Circuit Current (I_{SC}) 32

        2.4.3 Open-Circuit Voltage (V_{OC}) 32

        2.4.4. Maximum Power Point (MPP) 33

        2.4.5 Fill Factor (FF) 33

    2.5 PV Module Arrangement 33

        2.5.1 Typical Model of a Solar PV array 34

    2.6 Conclusion 35

CHAPTER III: SOLAR CELL TECHNOLOGIES 36

3.1 Introduction 36

3.2 Classification of solar Photovoltaic Cells 36

    3.2.1 Silicon Cells 37

    3.2.2 Single-Crystal Silicon Cells 37

    3.2.3 Multi-crystalline Silicon Cells 38

    3.2.4 Thin Silicon (Buried Contact) Cells 38

    3.2.5 Amorphous Silicon Cells 38

    3.2.6 Gallium Arsenide Cells 39

    3.2.7 Copper Indium (Gallium) Diselenide Cells 40

    3.2.8 Cadmium Telluride Cells 40

3.3 Emerging Technologies 41

3.4 Conclusion 42

CHAPTER IV: MAXIMUM POWER POINT TRACKING IN POWER CONDITIONING SYSTEM 43

4.1 Introduction 43

4.2 Need of MPPT Technique 43

    4.2.1 DC-AC Converter Systems 44

    4.2.2. DC-DC Converter Systems 45

4.3 Power Voltage and Voltage Current Characteristics of a Typical PV Module 46

4.4 Classification of MPPT Techniques 47

    4.4.1 P&O-MPPT Technique 47

    4.4.2 IC-MPPT Technique 50

4.5 Analysis of Other MPPT Techniques 51

    4.5.1 Fractional Open-Circuit Voltage Technique 52

    4.5.2. Fractional Short-Circuit Current Technique 52

4.6 Conclusion 52

CHAPTER V: INTRODUCTION TO POWER QUALITY 54

- 5.1 Introduction 54
- 5.2 Sources of Poor Power Quality 55
- 5.3 Need of Power Quality Concerns 55
- 5.4 Power Quality Problems 55
- 5.5 Solutions Adopted to Improve Power Quality 59
- 5.6 Power Quality Standards 59
- 5.7 Standards Related with Voltage Characteristics 59
    - 5.7.1. IEEE Standards 59
    - 5.7.2. IEC Electromagnetic Compatibility Standards 60
    - 5.7.3. The European Voltage Characteristics Standards 60
- 5.8 Standards Related With Current Characteristics 61
    - 5.8.1. IEEE Standards 61
    - 5.8.2. The International Electro Technical Commission 62
- 5.9 Conclusion 62

CHAPTER VI: MATLAB & SIMULINK SOFTWARE 63

- 6.1 Introduction 63
- 6.2 Introduction to Matlab Software 63
- 6.3 Matlab Windows 64
- 6.4 Matlab File Types 65
- 6.5 Simulink Software 65
    - 6.5.1 Controlling Execution of a Simulation 72
    - 6.5.2 Starting a Simulation 72
    - 6.5.3 Ending a Simulink Session 73
- 6.6 Conclusion 73

CHAPTER VII: SOLAR PV GRID CONNECTED SYSTEM WITH MPPT CONTROL .. 75
- 7.1 Introduction 75
- 7.2 System Computation Model 75
  - 7.2.1. Rotating Reference Frame Transformation 76
  - 7.2.2 Real & Reactive Power Control 77
  - 7.2.3. Synchronization and Control of Three-Phase Grid Connected Inverter System 79
  - 7.2.4. Generation of PWM Pulses 80
- 7.3 Controlling Scheme of Voltage Source Converter 82
- 7.4 Data MPPT Control for Maximum Power Point 84
- 7.6 Conclusion 89

CHAPTER VIII: EVALUATION AT DIFFERENT POWER FACTOR & FREQUENCY 91
- 8.1 Introduction 91
- 8.2 Simulation Results and Discussion 91
  - 8.2.1 Case I at different power factors and same frequency 91
  - 8.2.2. Case II at different frequency and same power factors 100
- 8.3 Conclusion 104

CHAPTER IX: MODEL ANALYSIS WITH INCREMENTAL CONDUCTANCE MPPT TECHNIQUE 106
- 9.1 Introduction 106
- 9.2 Voltage and Current Controllers 107
- 9.3 Solar PV Computation Model 108
- 9.4 Simulations Results and Discussion 110
- 9.5 Conclusion 115

CHAPTER X: MODEL ANALYSIS WITH PERTURB & OBSERVE MPPT TECHNIQUE 116
- 10.1 Introduction 116
- 10.2 Proposed System Configuration 116
- 10.3 Simulation Results and Case Studies 117
- 10.4 Conclusion 122

References 123

List of Abbreviations

Air Mass	(AM)
Balance of System	(BOS)
Basic Linear Algebra Subprograms	(BLAS)
Bipolar Junction Transistor	(BJT)
Building Integrated PV	(BIPV)
Clean Power Plan	(CPP)
Copper Indium Gallium Diselenide	(CIGS)
Conference of Parties	(COP)
Electromagnetic Compatibility	(EMC)
Energy Information Administration	(EIA)
European Nation	(EN)
Fast Fourier Transform	(FFT)
Gallium arsenide	(GaAs)
Gigawatts-Thermal	(GWth)
Incremental Conductance	(IC)
Individual Harmonic Distortion	(IHD)
Independent Power Producers	(IPP)
Institution of Electrical and Electronic Engineering	(IEEE)
International Energy Outlook	(IEO)
Intended Nationally Determined Contributions	(INDCs)
Integrated Development Environment	(IDE)
International Electro-technical Commission	(IEC)
kilo-watt-hour	(KWH)
Linear Algebra PACKage	(LAPACK)

MATrix LABoratory	(MATLAB)
Maximum Power Point	(MPP)
Maximum Power Point Tracking	(MPPT)
Metal Insulator Semiconductor Inversion Layer	(MIS-IL)
Metal Oxide Semiconductor Field Effect Transistor	(MOSFET)
Million Tonnes of Oil Equivalent	(MTOE)
Molybdenum	(Mo)
Organization for Economic Cooperation and Development	(OECD)
Perturb & Observe	(P&O)
Phase Locked Loop	(PLL)
Photovoltaic	(PV)
Power Quality	(PQ)
Power-Voltage	(P-V)
Power Conditioning System	(PCS)
Pulse Width Modulation	(PWM)
Root Mean Square	(rms)
Sine Pulse Width Modulation	(SPWM)
Standard Test Condition	(STC)
Sustainable Development Goal	(SDG)
Total Demand Distortion	(TDD)
Total Harmonic Distortion	(THD)
Transparent Conducting Oxide	(TCO)
United Nations Framework Convention on Climate Changes	(UNFCCC)
Voltage-Current	(V-I)
Voltage Source Converter	(VSC)
Watt-Peak	(Wp)

List of Tables

Table 1.1: Plan wise capacity addition in grid connected renewable capacity 27
Table 4.1: Algorithm of P&O-MPPT Technique 50
Table 4.2: Efficiency Comparison of P&O and IC-MPPT technique 51
Table 7.1: Data used at solar radiation S_x and temperatures T_x values 85
Table 7.2: System configuration parameters 85
Table 9.1: Parameters of PV grid connected system 109
Table 9.2: Specifications adopted for single PV array (Sun Power SPR-305-WHT) 109
Table 9.3: Total harmonic distortion analysis using IC-MPPT 115
Table 10.1: Total harmonic distortion analysis using P&O MPPT 122

List of Figures

Figure 1.1: World net electricity generation from renewable power (2012-2040) 23
Figure 2.1: Electrical circuit of an ideal PV cell 29
Figure 2.2: Single-diode exponential model of a PV cell 30
Figure 2.3: Downward shifting of the dark V-I curve (a) when light shines on a p-n junction diode and curve (b) is illuminated V-I curve 31
Figure 2.4: Voltage- Current and Power-Voltage solar cell characteristics 32
Figure 4.1: Block diagram of a solar PV grid connected application using MPPT 43
Figure 4.2: Typical connection scheme of a solar PV grid connected system 44
Figure 4.3: Changes in the V-I and P-V characteristics of the solar PV module due to change in radiation level 46
Figure 4.4: Changes in the V-I and P-V characteristics of the solar PV module due to change in temperature 47
Figure 4.5: Flowchart of Perturb & Observe MPPT technique 48
Figure 4.6 (a) Array P-V curve (b) Array V-I curve (c) Ramp up-down behavior of solar radiation intensity 49
Figure 4.7: Flowchart of Incremental conductance MPPT technique 51
Figure 5.1: Waveform for voltage sag 56
Figure 5.2: Waveform for voltage swell 56
Figure 5.3: Waveform for voltage interruption 57
Figure 5.4: Waveform for notching 58
Figure 5.5: Waveform for noise 58
Figure 6.1: Matlab desktop main menu and Simulink icon 66
Figure 6.2: Simulink library browser 67
Figure 6.3: Blocks in Sources sub-node 68
Figure 6.4: Blocks in Sinks sub-node 69
Figure 6.5: Drag and drop blocks to workspace from library browser 70
Figure 6.6: Components for a Simulink model 70
Figure 6.7: Complete model of Simulink connected 71
Figure 6.8: To simulate a Simulink model 72

Figure 6.9: Options of Simulink simulation..73
Figure 7.1 Simulink model of solar PV grid connected system...76
Figure 7.2: Power flow between the two AC sources 77
Figure 7.3: Generation of PWM pulses..81
Figure 7.4: Simulink control model of voltage source converter...83
Figure 7.5: Simulink active and reactive power measurements...84
Figure 7.6: (a) Solar radiation in W/m2 (b) Solar array current (c) Solar array voltage
 (d) Solar array power..87
Figure 7.7: actual DC link voltage and MPPT reference voltage 88
Figure 7.8: (a) Grid output power (b) Inverter output power...89
Figure 7.9: Modulation index...89
Figure 8.1: (a) Active power (b) Reactive power, variation of VSC, load and utility grid............92
Figure 8.2: FFT analysis of (a) VSC current and (b) grid current (Unity power factor, 50 Hz)....93
Figure 8.3: (a) Active power (b) Reactive power, variation of VSC, load and utility grid............94
Figure 8.4: FFT analysis of (a) VSC current and (b) grid current 95
Figure 8.5: (a) Active power (b) Reactive power, variation of VSC, load and utility grid
 (Power factor=0.707, 50 Hz)..96
Figure 8.6: FFT analysis of (a) VSC current and (b) grid current (Power factor=0.707, 50 Hz) ..97
Figure 8.7: (a) Active power (b) Reactive power, variation of VSC, load and utility grid
 (Power factor=0.50, 50 Hz)..98
Figure 8.8: FFT analysis of (a) VSC current and (b) grid current (Power factor=0.50, 50 Hz) 99
Figure 8.9: (a) Active power (b) Reactive power, variation of VSC, load and utility grid
 (Unity power factor, 49 Hz) 101
Figure 8.10: FFT analysis of (a) VSC current and (b) grid current (Unity power factor, 49 Hz) ..102
Figure 8.11: (a) Active power (b) Reactive power, variation of VSC, load and utility grid
 (Unity power factor, 51 Hz) 103
Figure 8.12: FFT analysis of (a) VSC current and (b) grid current (Unity power factor, 51 Hz) ..104
Figure 9.1: Simulink diagram of incremental conductance MPPT technique 106
Figure 9.2: Simulink diagram of boost converter control 107
Figure 9.3: Simulink diagram of DC voltage PI controller..107
Figure 9.4: Simulink diagram of DC current PI controller..107

Figure 9.5: Simulink diagram of solar PV with connected load 108
Figure 9.6: Different signal measurement blocks 110
Figure 9.7: Waveform of a) VSC Voltage b) VSC Current c) Load Voltage d) Load Current
    e) Utility grid Voltage f) Utility grid Current... 111
Figure 9.8: Change in actual DC link voltage with MPPT reference voltage.............................. 112
Figure 9.9: Modulation index... 112
Figure 9.10: Waveform of a) Real power and b) Reactive power, with IC MPPT technique 113
Figure 9.11: THD analysis of a) load current and b) grid current... 114
Figure 10.1: Simulink diagram of photovoltaic grid connected system 117
Figure 10.2: Waveform of a) VSC Voltage b) VSC Current c) Load Voltage
    d) Load Current e) Utility grid Voltage f) Utility grid Current.................................. 118
Figure 10.3: Waveform of a) Real power and b) Reactive power, with P&O MPPT technique ... 119
Figure 10.4: Change in actual DC link voltage with MPPT reference voltage............................ 120
Figure 10.5: Modulation index... 120
Figure 10.6: THD analysis of a) load current and b) grid current... 121

CHAPTER I: ELECTRIC POWER - A WORLD PERSPECTIVE

1.1. Introduction to Power Sector-A World Perspective

The worldwide mix of primary fuels used to generate electricity has changed a great deal over the past several decades. Coal continues to be the fuel most widely used in electricity generation [1], but there have been significant shifts to other generation fuels. Generation from nuclear power increased rapidly from the 1970s through the 1980s, and natural gas-fired generation increased considerably after the 1980s. The use of oil for generation declined after the late 1970s, when sharp increases in oil prices encouraged power generators to substitute other energy sources for oil.

Beginning in the early 2000s, concerns about the environmental consequences of greenhouse gas emissions heightened interest in the development of renewable energy sources, as well as natural gas-a fossil fuel that emits significantly less CO_2 than either oil or coal per Kilo-Watt-Hour (KWH) generated. In the International Energy Outlook (IEO) 2016, long-term global prospects continue to improve for generation from natural gas, nuclear, and renewable energy sources. Renewables are the fastest-growing source of energy for electricity generation, with annual increases averaging 2.9% from 2012 to 2040. In particular, non hydropower renewable resources are the fastest-growing energy sources for new generation capacity in both the Organization for Economic Cooperation and Development (OECD) and non-OECD regions. Non hydropower renewables accounted for 5% of total world electricity generation in 2012; their share in 2040 is 14% in the IEO-2016, with much of the growth coming from wind power.

After renewable energy sources, natural gas and nuclear power are the next fastest-growing sources of electricity generation. From 2012 to 2040, natural gas-fired electricity generation increases by 2.7%/year and nuclear power generation increases by 2.4%/year. With coal-fired generation growing by only 0.8%/year, renewable generation (including both hydropower and non hydropower resources) overtakes coal to become the world's largest source of energy for electricity generation by 2040. The outlook for coal-fired electricity generation could be further altered in the future by additional national policies or international agreements aimed at reducing or limiting its use. It should be noted that the IEO-2016 does not include implementation of the U.S. Clean Power Plan (CPP), which would reduce the use of coal in the United States substantially. Finally, if other nations with shale gas resources (notably, China) are able to replicate the U.S. success in exploiting shale gas production, the outlook for world natural gas-fired electricity generation could be much different from that represented in the IEO-2016.

1.1.1 Coal

Coal continues to be the largest single fuel used for electricity generation worldwide in the IEO-2016 until the end of the projection period, with renewable generation beginning to surpass coal-fired generation in 2040. Coal-fired generation, which accounted for 40% of total world electricity generation in 2012, declines to 29% of the total in 2040, despite a continued increase in total coal-fired electricity generation from 8.6 trillion kWh in 2012 to 9.7 trillion kWh in 2020 and 10.6 trillion kWh in 2040. Total electricity generation from coal in 2040 is 23% above the 2012 total.

1.1.2 Natural gas

Worldwide natural gas consumption for electricity generation grows in the IEO-2016 by an average of 2.7%/year from 2012 to 2040. From 22% of total world electricity generation in 2012, the natural gas share increases to 28% in 2040 in the IEO-2016. In the United States, natural gas-fired generation is encouraged by low prices and favorable greenhouse gas emission characteristics. Natural gas is the least carbon-intensive fossil fuel; like all fossil fuels, natural gas combustion emits carbon dioxide, but at about half the rate of coal.

1.1.3 Petroleum and Other Liquid Fuels

The use of petroleum and other liquid fuels for electricity generation continues to decline steadily in the IEO-2016. The share of total world generation from liquid fuels falls from 5% in 2012 to 2% in 2040, an average decline of 2.2%/year. Despite their recent decline, oil prices are expected to be higher in the long-term projection. As a result, liquids remain a more expensive option compared to other fuels used for generating electricity, and generators replace liquids-fired generation with other fuels where possible.

1.1.4 Renewable Resources

Renewables account for a rising share of the world's total electricity supply, and they are the fastest growing source of electricity generation in the IEO-2016 (Figure 1.1). Total generation from renewable resources increases by 2.9%/year, as the renewable share of world electricity generation grows from 22% in 2012 to 29% in 2040. Generation from non hydropower renewables is the predominant source of the increase, rising by an average of 5.7%/year and outpacing increases in natural gas (2.7%/year), nuclear (2.4%/year), and coal (0.8%/year), even without taking into account the growth in renewable generation anticipated under the CPP in the United States. By 2030, the CPP would increase U.S. renewables generation by roughly 396 billion kWh (58%) compared to the IEO-2016 Reference case, according to Energy Information Administration (EIA)'s analysis of the proposed CPP rule. Solar is the world's fastest-growing form of renewable energy, with net solar generation increasing by an average of 8.3%/year. Of the 5.9 trillion kWh of new renewable generation added over the projection period, hydroelectric and wind each account for 1.9 trillion kWh (33%), solar energy for 859 billion kWh (15%), and other renewables (mostly biomass and waste) for 856 billion kWh (14%).

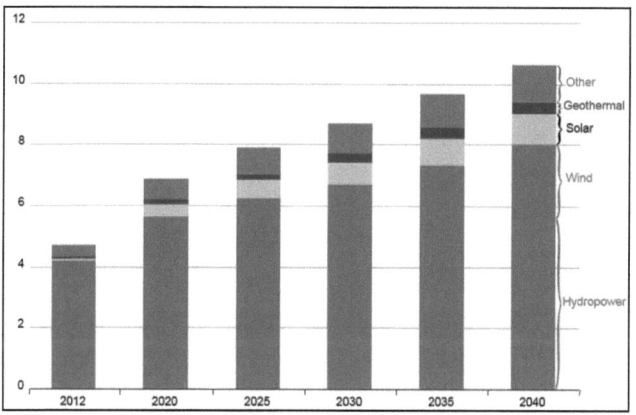

Figure 1.1: World net electricity generation from renewable power (2012-2040) [1]

1.2 Status of Indian Power Sector

Of the energy in India, the utility electricity sector had an installed capacity of 304.761 GW as of 31 July 2016 [2]. Renewable power plants constituted 28% of total installed capacity and Non-Renewable power plants constituted the remaining 72%. The gross electricity generated by utilities is 1,106 TWh (1,106,000 GWh) and 166 TWh by captive power plants during the 2014-15 fiscal. The gross electricity generation includes auxiliary power consumption of power generation plants. India became the world's third largest producer of electricity in the year 2013 with 4.8% global share in electricity generation surpassing Japan and Russia. During the year 2014-15, the per capita electricity generation in India was 1,010 kWh with total electricity consumption (utilities and non utilities) of 938.823 billion or 746 kWh per capita electricity consumption. Electric energy consumption in agriculture was recorded highest (18.45%) in 2014-15 among all countries. The per capita electricity consumption is lower compared to many countries despite cheaper electricity tariff in India.

Power is one of the most critical components of infrastructure crucial for the economic growth and welfare of nations. The existence and development of adequate infrastructure is essential for sustained growth of the Indian economy [3]. India's power sector is one of the most diversified in the world. Sources of power generation range from conventional sources such as coal, lignite, natural gas, oil, hydro and nuclear power to viable non-conventional sources such as wind, solar, and agricultural and domestic waste. Electricity demand in the country has increased rapidly and is expected to rise further in the years to come. In order to meet the increasing demand for electricity in the country, massive addition to the installed generating capacity is required. India ranks third, just behind US and China, among 40 countries with renewable energy focus, on back of strong focus by the government on promoting renewable energy and implementation of projects in a time bound manner.

Indian power sector is undergoing a significant change that has redefined the industry outlook. Sustained economic growth continues to drive electricity demand in India. The Government of India's focus on attaining 'Power for all' has accelerated capacity addition in the country. At the same time, the competitive intensity is increasing at both the market and supply sides (fuel, logistics, finances, and manpower). Total capacity of renewable energy plants in India stood at 42,850 MW as on April 30, 2016, thereby surpassing the 42,783 MW capacity of large hydroelectricity projects in the country. Cumulative solar installations in India crossed the 7.5 GW mark in May 2016, about 2.2 GW more than all of the solar installations in 2015.

The Indian Planning Commission's 12th five-year plan estimates total domestic energy production to reach 669.6 Million Tonnes of Oil Equivalent (MTOE) by 2016–17 and 844 MTOE by 2021-22. As of January 2016, total thermal installed capacity stood at 200.74 GW, while hydro (renewable) energy installed capacity totaled 42.66 GW. At 5.78 GW, nuclear energy capacity remained broadly constant compared with the previous year. India's rooftop solar capacity addition grew 66 per cent from last year to reach 525 MW, and has the potential to grow up to 6.5 GW. India's wind power capacity, installed in financial year-2016, is estimated to increase 20% over last year to 2,800 MW, led by favorable policy support that has encouraged both Independent Power Producers (IPP) and non-IPPs. India is expected to add nearly 4,000 MW of solar power in 2016, nearly twice the addition of 2,133 MW in 2015.

India's wind energy market is expected to attract investments totaling Rs 1,00,000 crore (US$ 14.82 billion) by 2020, and wind power capacity is estimated to almost double by 2020 from over 23,000 MW in June 2015, with an addition of about 4,000 MW per annum in the next five years.

1.3 Status of Global Renewable Energy

The year 2015 was an extraordinary one for renewable energy, with the largest global capacity additions seen to date, although challenges remain, particularly beyond the power sector [4]. The year saw several developments that all have a bearing on renewable energy, including a dramatic decline in global fossil fuel prices; a series of announcements regarding the lowest-ever prices for renewable power long-term contracts; a significant increase in attention to energy storage; and a historic climate agreement in Paris that brought together the global community.

Renewables are now established around the world as main-stream sources of energy. Rapid growth, particularly in the power sector, is driven by several factors, including the improving cost-competitiveness of renewable technologies, dedicated policy initiatives, better access to financing, energy security and environmental concerns, growing demand for energy in developing and emerging economies, and the need for access to modern energy. Consequently, new markets for both centralized and distributed renewable energy are emerging in all regions. 2015 was a year of firsts and high-profile agreements and announcements related to renewable energy. These include commitments by both the G7 and the G20 to accelerate access to renewable energy and to advance energy efficiency, and the United Nations General Assembly's adoption of a dedicated Sustainable Development Goal on Sustainable Energy for All (SDG 7).

The year's events culminated in December at the United Nations Framework Convention on Climate Change's (UNFCCC) 21st Conference of the Parties (COP21) in Paris, where 195 countries agreed to limit global warming to well below 2 degrees Celsius. A majority of countries committed to scaling up renewable energy and energy efficiency through their Intended Nationally Determined Contributions (INDCs). Out of the 189 countries that submitted INDCs, 147 countries mentioned renewable energy, and 167 countries mentioned energy efficiency; in addition, some countries committed to reforming their subsidies for fossil fuels. Precedent-setting commitments to renewable energy also were made by regional, state and local governments as well as by the private sector.

Although many of the initiatives announced in Paris and elsewhere did not start to affect renewable markets in 2015, there already were signs that a global energy transition is under way. Renewable energy provided an estimated 19.2% of global final energy consumption in 2014, and growth in capacity and generation continued in 2015. An estimated 147 GW of renewable power capacity was added in 2015, the largest annual increase ever, while renewable heat capacity increased by around 38 Gigawatts-Thermal (GWth), and total biofuels production also rose. This growth occurred despite tumbling global prices for all fossil fuels, ongoing fossil fuel subsidies and other challenges facing renewables, including the integration of rising shares of renewable generation, policy and political instability, regulatory barriers and fiscal constraints. Global investment also climbed to a new record level, in spite of the plunge in fossil fuel prices, the strength of the US dollar (which reduced the dollar value of non-dollar investments), the continued weakness of the European economy and further declines in per unit costs of wind and solar Photovoltaic (PV). For the sixth consecutive year, renewables outpaced fossil fuels for net investment in power capacity additions.

Private investors stepped up their commitments to renewable energy significantly during 2015. The year witnessed both an increase in the number of large banks active in the renewables sector and an increase in loan size, with major new commitments from international investment firms to renewables and energy efficiency. New investment vehicles-including green bonds, crowd funding and yield cos-expanded during the year. Mainstream financing and securitization structures also continued to move into developing country markets as companies (particularly solar PV) and investors sought higher yield, even at the expense of higher risk. In parallel with growth in markets and investments, 2015 saw continued advances in renewable energy technologies, ongoing energy efficiency improvements, increased use of smart grid technologies and significant progress in hardware and software to support the integration of renewable energy, as well as progress in energy storage development and commercialization. The year also saw expanded use of heat pumps, which can be an energy-efficient solution for heating and cooling.

Employment in the renewable energy sector (not including large-scale hydropower) increased in 2015 to an estimated 8.1 million jobs (direct and indirect). Solar PV and bio-fuels provided the largest numbers of renewable energy jobs. Large-scale hydropower accounted for an additional 1.3 million direct jobs. Considering all renewable energy technologies, the leading employers in 2015 were China, Brazil, the United States and India. India has a vast supply of renewable energy resources, and it has one of the largest programs in the world for developing renewable energy based product and systems. India is a developing and fast growing large economy and faces a great challenge to meet its energy needs in a responsible and sustainable manner. Grid-interactive renewable power capacity in the country reached 44244.27 MW on 30th June 2016 (Table 1.1), which is about 14.5% of the total grid installed capacity in the country. Its contribution is about 8% to total electric power generation.

Table 1.1: Plan wise capacity addition in grid connected renewable capacity [5]

Resource	Financial year 2016-2017		Cumulative Achievements
	Target (MW)	Achievement (MW)	(as on 30.06.2016) (MW)
Wind	4000	373.95	27151.40
Small-hydro	250	30.30	4304.25
Biomass	400	29.50	4860.83
Waste	10	7.5	122.58
PV	12000	1042.35	7805.21
Total	16660	1483.60	44244.27

1.4 Conclusion

This chapter mainly presents the world perspective and growth of electric power through conventional sources and its vis-à-vis comparative growth with various sources of renewable energy. It has been discussed that electric power has been one of the most critical component among others available required for infrastructure and crucial for economic growth and welfare of any nation. The growth of any developing nation mainly depends on various resources critically required in growth of infrastructure. This chapter also highlights the recent advancements recently taken place in renewable energy. The measures being adopted for building and supporting in the further renewable energy growth in developing and developed economies is also presented.

CHAPTER II: AN INTRODUCTION TO SOLAR PHOTOVOLTAIC CELL & MODELING

2.1 Introduction

This chapter mainly the fundamentals on solar PV cells with the help of mathematical analysis. Basic output current and voltage for a single PV cell has been presented and its importance is described. In addition, the dependence of a solar PV cell on variable solar radiation and ambient temperature is also discussed. This discussion has been carried out using Power-Voltage (*P-V*) and Voltage-Current (*V-I*) curves at Maximum Power Point (MPP). Also, few models on a solar PV cell based on empirical equations have been elaborated.

2.2 A Solar PV Cell-A p-n Junction Diode

A solar PV cell is a semiconductor device, which behaves as a current source when driven by a flux of solar radiation from the sun. This occurs, when radiation is incident upon absorbing material and separates positive and negative charge carriers in the presence of an electric field. The electric field exists permanently at junctions or in homogeneities in solar cells, which can be described as silicon semiconductor junction device. A silicon semiconductor junction device contains a p-n junction similar to that of a common diode; however in a solar cell, it exists over a large surface area. When solar PV cell is not illuminated and connected to a forward bias, it mimics the electrical characteristics of an ideal diode, modeled by Equation 2.1, where the current produced is referred to as the dark current I_d:

$$I_d = I_{rs}(e^{\frac{qV_a}{kT_c}} - 1) \qquad (2.1)$$

where, I_{rs} is reverse saturation current of the diode (A), q is electron charge (1.602 × 10^{-19} C), V_a is the PV cell output voltage (V), k is Boltzmann's constant (1.38 × 10^{-23} J/K) and T_c is the solar cell temperature (°C). The current in the cell that results from solar radiation is called the photo current, which flows in the direction opposite of the forward dark current. Its value remains the same regardless of external voltage and therefore, it can be measured by the short-circuit current. This current varies linearly with the intensity of solar radiation as a large radiation is able to separate more number of charge carriers. The overall current is then described as the difference between the dark current and the photocurrent [6-7]. The current I_a produced by an illuminated cell can be written as shown in Equation 2.2,

$$I_a = I_{ph} - I_{rs}(e^{\frac{qV_a}{kT_c}} - 1) \qquad (2.2)$$

This is the mathematical equation, which models the behavior of an ideal PV cell shown in Figure 2.1. The production of photocurrent is modeled with an ideal current source and the dark current is modeled with a diode referred to as the diffusion diode, which is represented by a single-diode model. Some authors have proposed more sophisticated models, [8-9], in which an extra diode is used to represent the effect of the recombination of carriers. However, for simplicity, we have studied and used the single-diode model of solar PV cell. The single-diode model offers good compromise between the accuracy and simplicity. The simplicity of this model makes the model easy and effective when used for the simulation of PV devices with power converters.

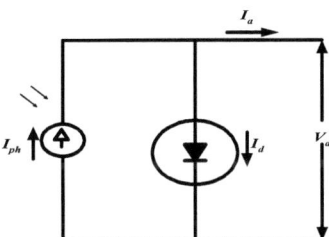

Figure 2.1: Electrical circuit of an ideal PV cell

Figure 2.2 depicts a well-known equivalent circuit of single solar cell composed of a light generated current source, a single-diode representing the non-linear impedance of p-n junction with series resistance R_s and shunt resistance R_p. The series resistance accounts for any resistance in current path through semiconductor material, metal grid, contacts, and current collecting bus [10]. The shunt resistance as mentioned in reference [11] is a loss associated with a slight leakage current through a parallel resistive path to the device. It is not noticeable as series resistance, because the effects are minimal unless a number of PV modules are connected in parallel for a large system.

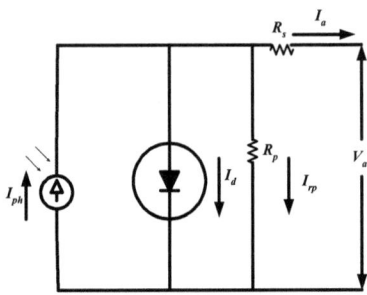

Figure 2.2: Single-diode exponential model of a PV cell

2.3 Difference Between a Solar PV Cell and a p-n Junction Diode

In a p-n junction diode, there are four current components, which are present in equilibrium conditions, namely, electron drift, electron diffusion, hole drift and hole diffusion. In equilibrium condition, the net current is zero which requires drift and diffusion current of the carriers to be equal and opposite. When light shines on the solar PV cells, it results in a large drift current due to minority electrons and holes. These carriers flow from n-side to p-side. Since this current flow is caused by the incident light, it is known as light generated current or photocurrent I_{ph}. Note that the generated photo-voltage due to light biases the p-n junction in a forward biased mode i.e. the generated photo-voltage reduces the junction potential energy barrier. As a result, there is a diffusion current, which flows in the opposite direction of photocurrent. But the magnitude of photocurrent is larger than the forward biased diffusion current and hence, the net current flows from the n-side to p-side (opposite to that of forward biased diode current). Thus, when the light shines on a PV cell, the current flows in the opposite direction to that of the generated voltage. Overall, the effect of light shining is to shift the V-I curve of the diode downwards on the current voltage axis, as depicted in the Figure 2.3. It is found that in the fourth quadrant of this curve, the voltage is positive and current is negative resulting in negative power. The negative power implies that the power can be extracted from the device. Therefore, a solar PV cell generates the power, instead of consuming power like the other electronic devices, where the power is positive.

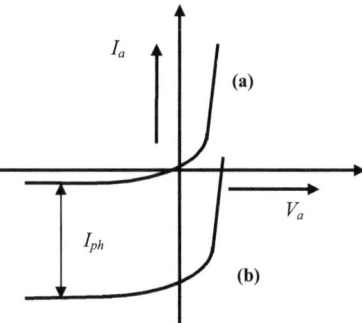

Figure 2.3: Downward shifting of the dark V-I curve (a) when light shines on a p-n junction diode and curve (b) is illuminated V-I curve [12]

2.4 Basic Parameters of Solar PV Cell

Normally, the electrical characteristics of a PV cell are displayed as a relation between the cell voltage and current, and a relation between the cell voltage and power. However, several electric quantities are important to describe the operation of the PV cell. These electric quantities include: the cell voltage under open-circuit conditions (V_{OC}), the cell current under short-circuit conditions (I_{SC}), the cell voltage, current and power at MPP, V_{MPP}, I_{MPP}, and, P_{MPP}, respectively.

2.4.1 Voltage-Current (V-I) Characteristics

V-I characteristics curve represents all possible current and voltage operating points for PV cell. This curve can be generated by changing the electrical load value during lab experiment. As can be seen in Figure 2.4, when voltage increases, the current start at its maximum value then decreases gradually to reach the zero. The operating point at *V-I* curve is determined by the electrical load. Certain points at the *V-I* curve are highlighted to rate the PV module performance and are used for the design of PV systems. These measured values are taken at the Standard Test Condition (STC), which are at solar radiation ($G=1000$ W/m^2), Temperature ($T= 25°$ C), and Air Mass (AM=1.5). The knee points, in both *V-I* and *P-V* curves, represent the MPP. The optimum electrical load is the load that operates the PV at its MPP, if the PV generator is able to deliver maximum power.

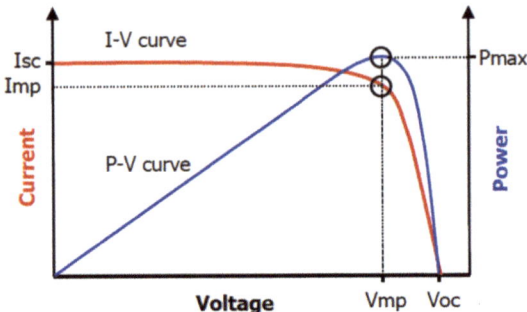

Figure 2.4: Voltage- Current and Power-Voltage solar cell characteristics

The electrical characteristics of a PV cell depend on the variable solar radiation and cell temperature. There is a proportional relationship between the current and the solar radiation at different levels of the solar radiation and constant temperature. Any change in solar radiation has a strong effect on the short-circuit current and the output power of the solar cell, but negligible effect on the open-circuit voltage. On the other hand, any change in the temperature at constant solar radiation has a significant effect on the open-circuit voltage and output power of the cell, while negligible effect on the short-circuit current.

2.4.2 Short Circuit Current (I_{SC})

Short-circuit current is the maximum possible current in the circuit, at zero voltage as shown in Figure 2.4. In other words, when there is negligible electrical load in the circuit. This case occurs by connecting the positive and negative terminals together. The short-circuit current is directly proportional to the available sunlight.

2.4.3 Open-Circuit Voltage (V_{OC})

Open-circuit voltage is the maximum voltage, at zero current. In other words, open-circuit voltage represents the maximum possible voltage value, which occurs when a huge load in connected to the circuit or in case of no load.

2.4.4. Maximum Power Point (MPP)

In case of short-circuit or open-circuit operating points, there is no power generated from a solar PV cell. As a result, the operating point should fall into the range of the maximum power output of PV cell. This operating point is determined by choosing the correct value of the connected load. The maximum output power is defined as the multiplication of the voltage and the current at the MPP. Where V_{MPP} and I_{MPP} are voltage and current the values that gives the maximum operating power. MPP is the rate of the unit by W_p (Watt-Peak).

2.4.5 Fill Factor (FF)

The fill factor is an indicator of the quality of the PV cell. The sharpness of the knee in an V-I curve indicates how well a p-n junction is manufactured. The maximum value of the fill factor is one, which is theoretical value only. The maximum practical value in silicon is 0.88. FF is defined as the ratio between the maximum generated power at MPP and the maximum theoretical power (I_{sc} multiply by V_{oc}).

For a good V-I curve profile with fill factor close to unity, the current should stay constant while the voltage increases till the operating point reach the knee of the curve. After that point, the current of a solar PV cell should also gradually decrease.

2.5 PV Module Arrangement

PV modules are composed of series and parallel connection of solar PV cells with blocking and bypass diodes as additional components. While the manufacture and size of solar PV cells vary, in general, a single PV cell has a relatively low voltage handling capability on the order of 0.6 V. In order to package PV cells as a more practical device, most manufacturers produce solar modules. In the modules, a group of solar PV cells are connected in series and parallel in order to increase the voltage and current handling capability, respectively.

The major factors that affect the efficiency of a solar PV module are its ambient temperature and solar radiation. The output voltage of a solar PV cell is a function of load current and depends on the photocurrent that is determined by the incident solar radiation level during operation, [13]. The V-I equation of a module is similar to that of solar PV cell. The V-I curve of the module is the combination of V-I curve of all solar cells connected in a module. Equation 2.3 gives output current I_a [13-14] for a PV array as:

$$I_a = I_{ph}N_p - I_{rs}N_p e^{\{(\frac{q}{akT_c}(\frac{V_a}{N_s}+\frac{I_a R_s}{N_p})-1)\}} - \frac{N_p}{R_p}(\frac{V_a}{N_s}+\frac{I_a R_s}{N_p}) \quad (2.3)$$

where,
$$I_{ph} = [I_{sc} + K_i(T_c - T_x)]\frac{S_c}{S_x}$$

The curve fitting factor or identity factor a is used to adjust the V-I curve of the array. The output voltage of a single solar PV cell is very small for most applications. Therefore, to produce useful DC voltage, a number of solar PV cells are connected in series and mounted on a support frame, which forms a solar PV module (or a solar PV array panel). If the shunt resistance R_p is assumed as infinity and $N_s=N_p=1$, Equation 2.3 becomes

$$I_a = I_{ph} - I_{rs} e^{\{(\frac{q}{akT_c}(V_a+I_a R_s)-1)\}}$$

2.5.1 Typical Model of a Solar PV array

In a similar way, the V-I characteristics of a PV array will be similar to that of PV module, except that it is the combination of several V-I equations of PV modules. The PV cell output voltage Equation 2.4 is a function of the photocurrent that is mainly determined by load current, solar radiation level and operating cell temperature during operation.

$$V_a = \frac{akT_c}{q} \ln(\frac{I_{ph}+I_{rs}-I_a}{I_{rs}}) - R_s I_a \quad (2.4)$$

The above gives output voltage of a single solar cell, which is multiplied by number of cells connected in series gives full array voltage. If temperature and solar radiation level changes, the voltage and current output values of PV array also follows this change. The effects of change in temperature and solar radiation levels have also been included in final PV array model. According to the reference [13], for a known temperature and a known solar radiation level, a model is obtained which is then modified to handle different cases of temperature and solar radiation levels. The operating temperature of the solar cell varies as a function of solar radiation level, and ambient temperature. The variable ambient temperature affects the cell output voltage, and cell photocurrent. These effects are represented in the model [13-14] by Equation 2.5 and Equation 2.6 as,

$$C_{tv} = 1 + \beta_t(T_a - T_x) \qquad (2.5)$$

$$C_{ti} = 1 + \gamma \frac{(T_x - T_a)}{S_c} \qquad (2.6)$$

where, $\beta_t = 0.0042$ and $\gamma = 0.062$ for the cell used and $T_a = 20\,°\text{C}$, is ambient temperature during the cell testing. This is used to obtain the modified model of cell for another ambient temperature T_x. A change in solar radiation causes a change in cell photocurrent, and operating temperature, which in turn affects cell output voltage. If solar radiation increases from S_{x1} to S_{x2}, cell operating temperature and photocurrent will also increase from T_{x1} to T_{x2} and from I_{ph1} to I_{ph2}, respectively. Thus change in operating temperature and photocurrent due to variation in solar radiation are expressed as,

$$C_{sv} = 1 + \beta_t \alpha_s (S_x - S_c) \qquad (2.7)$$

$$C_{si} = 1 + \frac{(S_x - S_c)}{S_c} \qquad (2.8)$$

Here, S_c is benchmark reference solar radiation level during cell testing to obtain modified cell model. S_x is new level of solar radiation. The constant α_s represents slope of change in cell operating temperature due to change in solar radiation level and is equal to 0.011 for used solar cells. Using Equation 2.5 to Equation 2.8, the new values of cell output voltage V_{an} and photocurrent I_{phn} at new temperature T_x and solar radiation S_x are obtained as,

$$V_{an} = C_{tv} C_{sv} V_a \qquad (2.9)$$

$$I_{phn} = C_{ti} C_{si} I_{ph} \qquad (2.10)$$

V_a and I_{ph} are benchmark reference cell output voltage and reference cell photocurrent, respectively.

2.6 Conclusion

In this chapter, an introduction of a solar PV cell with its behavior with the help of its electrical circuits in ideal and practical conditions is presented. This description is carried out using its basic output current and output voltage equations. In addition, the various parameters as defined in a standard manufacturer data-sheet is discussed. The basic difference between a solar PV cell and a normal p-n junction diode is also described. The importance of MPP using V-I and P-V electrical characteristics is highlighted. In order to study any PV model, the dependence of a solar PV cell on variable solar radiation and ambient temperature is derived and discussed.

CHAPTER III: SOLAR CELL TECHNOLOGIES

3.1 Introduction

Traditional solar cells are made from silicon, and are generally the most efficient. The solar cell technologies based on silicon wafer are generally referred as the first generation technologies, while the cell technologies based on thin film are referred as the second generation technologies. Thin-film solar cells made from amorphous silicon or non-silicon materials such as cadmium telluride are the second-generation solar cells, and are gaining a greater share in overall installations. Since the initial development of reasonable solar cell efficiency of 6% in 1954, silicon has been a dominating material [12]. As of today, more than 90% of solar cells are realized in silicon wafers, either mono-crystalline or multi-crystalline. It was possible because silicon provides reasonable high and stable solar cell efficiencies. The use of silicon in PV industry also gains from the enormous knowledge that the microelectronic industry has created on silicon, starting from wafer manufacturing to the solar cell fabrication technology.

This chapter focuses and covers thin film technologies such as amorphous silicon, CdTe, CIGS and thin film crystalline silicon solar cell technologies. The primary objective of development of thin film technologies is to reduce the cost of PV modules significantly lower than the cost of PV modules obtained from wafer-based solar PV modules.

3.2 Classification of solar Photovoltaic Cells

The general classification of solar PV cells is presented below:

- Silicon Cells
- Single-Crystal Silicon Cells
- Multi-crystalline Silicon Cells
- Thin Silicon (Buried Contact) Cells
- Amorphous Silicon Cells
- Gallium Arsenide Cells
- Copper Indium (Gallium) Diselenide Cells
- Cadmium Telluride Cells

3.2.1 Silicon Cells

Silicon solar PV cells come in several varieties. The most common cell is the single-crystal silicon cell. Other variations include multi-crystalline (poly-crystalline), thin silicon (buried contact) cells, and amorphous silicon cells.

3.2.2 Single-Crystal Silicon Cells

While single-crystal silicon cells are still the most common cells, the fabrication process of these cells is relatively energy intensive, resulting in limits to cost reduction for these cells. Since single-crystal silicon is an indirect band-gap semiconductor (energy gap, E_g=1.1 eV), its absorption constant is smaller than that of direct band-gap materials. This means that single-crystal silicon cells need to be thicker than other cells in order to absorb a sufficient percentage of incident solar radiation. This results in the need for more material and correspondingly more energy involved in cell processing, especially since the cells are still produced mostly by sawing of single-crystal silicon ingots into wafers that are about 200 μm thick. To achieve maximum fill of the module, round ingots are first sawed to achieve closer to a square cross-section prior to wafering.

After chemical etching to repair surface damage from sawing, the junction is diffused into the wafers. Improved cell efficiency can then be achieved by using a preferential etch on the cell surfaces to produce textured surfaces. The textured surfaces reflect photons back toward the junction at an angle, thus increasing the path length and increasing the probability of the photon being absorbed within a minority carrier diffusion length of the junction. Following the chemical etch, contacts, usually aluminum, are evaporated and annealed and the front surface is covered with an antireflective coating.

The cells are then assembled into modules, consisting of approximately 33 to 36 individual cells connected in series. Since the open-circuit output voltage of an individual silicon cell typically ranges from 0.5 V to 0.6 V, depending upon solar radiation level and cell temperature, this results in a module open-circuit voltage between 18 V and 21.6 V. The cell current is directly proportional to the solar radiation and the cell area. A 4-ft^2 (0.372-m^2) module (active cell area) under full sun will typically produce a maximum power close to 55 W at approximately 17 V and 3.2 A.

3.2.3 Multi-crystalline Silicon Cells

By pouring molten silicon into a crucible and controlling the cooling rate, it is possible to grow multi-crystalline silicon with a rectangular cross-section. This eliminates the "squaring-up" process and the associated loss of material. The ingot must still be sawed into wafers, but the resulting wafers completely fill the module. The remaining processing follows the steps of single-crystal silicon, and cell efficiencies in excess of 15% have been achieved for relatively large area cells. Multi-crystalline material still maintains the basic properties of single-crystal silicon, including the indirect band-gap. Hence, relatively thick cells with textured surfaces have the highest conversion efficiencies. Multi-crystalline silicon modules are commercially available and are recognized by their speckled surface appearance.

3.2.4 Thin Silicon (Buried Contact) Cells

The current flow direction in most PV cells is between the front surface and the back surface. In the thin silicon cell, a dielectric layer is deposited on an insulating substrate, followed by alternating layers of n-type and p-type silicon, forming multiple p-n junctions. Channels are then cut with lasers and contacts are buried in the channels, so the current flow is parallel to the cell surfaces in multiple parallel conduction paths. These cells minimize resistance from junction to contact with the multiple parallel conduction paths and minimize blocking of incident radiation by the front contact. Although the material is not single-crystal, grain boundaries cause minimal degradation of cell efficiency. The collection efficiency is very high, since essentially all photon-generated carriers are generated within a diffusion length of a p-n junction. This technology is relatively new, but has already been licensed to a number of firms worldwide [15].

3.2.5 Amorphous Silicon Cells

Amorphous silicon has no predictable crystal structure. As a result, the uniform covalent bond structure of single-crystal silicon is replaced with a random bonding pattern with many open covalent bonds. These bonds significantly degrade the performance of amorphous silicon by reducing carrier mobility and the corresponding diffusion lengths. However, if hydrogen is introduced into the material, its electron will pair up with the dangling bonds of the silicon, thus passivating the material. The result is a direct band-gap material with a relatively high absorption constant. A film with a thickness of a few micrometers will absorb nearly all incident photons with energies higher than the 1.75 eV band-gap energy. Maximum collection efficiency for a-Si:H is achieved by fabricating the cell with a pin junction. Early work on the cells revealed, however, that if the intrinsic region is too thick, cell performance will degrade over time. This problem has now been overcome by the manufacture of multi-layer cells with thinner pin junctions. In fact, it is possible to further increase cell efficiency by stacking cells of a-SiC:H on top, a-Si:H in the center, and a-SiGe:H on the bottom. Each successive layer from the top has a smaller bandgap, so the high-energy photons can be captured soon after entering the material, followed by middle-energy photons and then lower energy photons. While the theoretical maximum efficiency of a-Si:H is 27% [16], small-area lab cells have been fabricated with efficiencies of 14% and large-scale devices have efficiencies in the 10% range [17]. Amorphous silicon cells have been adapted to the Building Integrated PV (BIPV) market by fabricating the cells on stainless steel and polymide substrates. The solar shingle is now commercially available, and amorphous silicon cells are commonly used in solar calculators and solar watches.

3.2.6 Gallium Arsenide Cells

Gallium arsenide (GaAs), with its 1.43 eV direct band-gap, is a nearly optimal PV cell material. The only problem is that it is very costly to fabricate cells. GaAs cells have been fabricated with conversion efficiencies above 30% and with their relative insensitivity to severe temperature cycling and radiation exposure, they are the preferred material for extraterrestrial applications, where performance and weight are the dominating factors. Gallium and arsenic react exothermically when combined, so formation of the host material is more complicated than formation of pure, single-crystal silicon. Modern GaAs cells are generally fabricated by growth of a GaAs film on a suitable substrate, such as Ge. A typical GaAs cell has a Ge substrate with a layer of n-GaAs followed by a layer of p-GaAs and then a thin layer of p-GaAlAs between the p-GaAs and the top contacts. The p-GaAlAs has a wider bandgap (1.8 eV) than the GaAs, so the higher energy photons are not absorbed at the surface, but are transmitted through to the GaAs pn junction, where they are then absorbed. Recent advances in III-V technology have produced tandem cells similar to the a-Si:H tandem cell. One cell consists of two tandem GaAs cells, separated by thin tunnel junctions of GaInP, followed by a third tandem GaInP cell, separated by AlInP tunnel junctions. The tunnel junctions mitigate voltage drop of the otherwise forward-biased p-n junction that would appear between any two tandem p-n junctions in opposition to the photon-induced cell voltage.

3.2.7 Copper Indium (Gallium) Diselenide Cells

Another promising thin film material is Copper Indium Gallium Diselenide (CIGS). While the basic copper indium diselenide cell has a bandgap of 1.0 eV, the addition of gallium increases the band-gap to closer to 1.4 eV, resulting in more efficient collection of photons near the peak of the solar spectrum. CIGS has a high absorption constant and essentially all incident photons are absorbed within a distance of 2 μm, as in a-Si:H. Indium is the most difficult component to obtain, but the quantity needed for a module is relatively minimal. The CIGS cell is fabricated on a soda glass substrate by first applying a thin layer of Molybdenum (Mo) as the back contact, since the CIGS will form an ohmic contact with Mo. The next layer is p-type CIGS, followed by a layer of n-type CdS, rather than n-type CIGS, because the p-n homo-junction in CIGS is neither stable nor efficient. While the cells discussed thus far have required metals to obtain ohmic front contacts, it is possible to obtain an ohmic contact on CdS with a Transparent Conducting Oxide (TCO) such as ZnO. The top surface is first passivated with a thin layer (50 nm) of intrinsic ZnO to prevent minority carrier surface recombination. Then a thicker layer (350 nm) of n+ ZnO is added, followed by an MgF_2 antireflective coating.

Efficiencies of laboratory cells are now near 18%, with a module efficiency of 11.1% have already been reported. CIGS modules were not commercially available, the technology has been under field tests since last 15 years. It has been projected that the cells may be manufactured on a large scale for $1/W or less. At this cost level, area related costs become significant, so that it becomes important to increase cell efficiency to maximize power output for a given cell area.

3.2.8 Cadmium Telluride Cells

Of the II-VI semiconductor materials, CdTe has a theoretical maximum efficiency of near 25%. The material has a favorable direct band-gap (1.44 eV) and a large absorption constant. As in the other thin film materials, a 2-μm thickness is adequate for the absorption of most of the incident photons. Small laboratory cells have been fabricated with efficiencies near 15% and module efficiencies close to 10% have been achieved. Some concern has been expressed about the Cd content of the cells, particularly in the event of fire dispersing the Cd. It has been determined that anyone endangered by Cd in a fire would be far more endangered by the fire itself, due to the small quantity of Cd in the cells. Decommissioning of the module has also been analyzed and it has been concluded that the cost to recycle module components is pennies per watt. The CdTe cell is fabricated on a glass superstrate covered with a thin Transparent Conducting Oxide (TCO) (1 µm). The next layer is n-type CdS with a thickness of approximately 100 nm, followed by a 2-µm thick CdTe layer and a back contact of an appropriate metal for ohmic contact, such as Au, Cu/Au, Ni, Ni/Al, ZnTe:Cu or (Cu, HgTe). The back contact is then covered with a layer of ethylene vinyl acetate (EVA) or other suitable encapsulant and another layer of glass. The front glass is coated with an antireflective coating. Experimental CdTe arrays up to 25 kW have been under test for several years with no reports of degradation. It has been estimated that the cost for large-scale production can be reduced to below $1/W. Once again, as in the CIGS case, module efficiency needs to be increased to reduce the area-related costs.

3.3 Emerging Technologies

The PV field is moving so quickly that by the time information appears in print, it is generally outdated. Reliability of cells, modules, and system components continues to improve. Efficiencies of cells and modules continue to increase, and new materials and cell fabrication techniques continue to evolve. One might think that Si cells will soon become historical artifacts. This may not be the case. Efforts are underway to produce Si cells that have good charge carrier transport properties while improving photon absorption and reducing the energy for cell production. Ceramic and graphite substrates have been used with thinner layers of Si. Processing steps have been doubled up. Metal Insulator Semiconductor Inversion Layer (MIS-IL) cells have been produced in which the diffused junction is replaced with a Schottky junction. By use of clever geometry of the back electrode to reduce the rear surface recombination velocity along with front surface passivation, an efficiency of 18.5% has been achieved for a laboratory MIS-IL cell. Research continues on ribbon growth in an effort to eliminate wafering, and combining crystalline and amorphous Si in a tandem cell to take advantage of the two different band-gaps for increasing photon collection efficiency has been investigated. At least eight different CIS-based materials have been proposed for cells. The materials have direct band-gaps ranging from 1.05 to 2.56 eV. A number of III-V materials have also emerged that have favorable photon absorption properties. In addition, quantum well cells have been proposed that have theoretical efficiencies in excess of 40% under concentrating conditions. The solar PV market seems to have taken a strong foothold, with the likelihood that annual solar PV module shipments will exceed 200 MW before the end of the century and continue to increase by approximately 15% annually as new markets open as cost continues to decline and reliability continues to improve.

3.4 Conclusion

This chapter has presented a discussion on advantages and limitations of crystalline and thin film technologies with their common features. This chapter has begun from by looking at the development of solar cells from the early stage. Then the commercial silicon wafer based solar cell technology is described. Special attention has been given to each thin film technology as its related material properties, fabrication techniques are discussed. The chapter has covered thin film technologies such as amorphous silicon, CdTe, CIGS, and thin film crystalline silicon solar cell technologies.

CHAPTER IV: MAXIMUM POWER POINT TRACKING IN POWER CONDITIONING SYSTEM

4.1 Introduction

When a solar PV system is used in a system, its operating point is decided by the load to which it is connected [12]. Also, since solar radiation falling on a PV module varies throughout the day, the operating point of module also changes throughout the day. In order to ensure the operation of PV modules for maximum power transfer, a special method called Maximum Power Point Tracking (MPPT) is employed in PV systems, which is explained in the following paragraphs. In case of MPPT, electronic circuitry is used to ensure that maximum amount of generated power is transferred to the load.

4.2 Need of MPPT Technique

A MPPT technique is used for extracting the maximum power from the solar PV module and transferring that power to the load. A DC-DC converter (step up/or step down) algorithm and inverter DC-AC serves the purpose of transferring maximum power from the solar PV module to the connected load. A DC-DC converter and DC/AC inverter acts as an interface between the load and module as shown in Figure 4.1. By changing the duty cycle of the DC-DC converter, the load impedance as seen by the source is varied and matched at the point of the peak power with the source to transfer the maximum power.

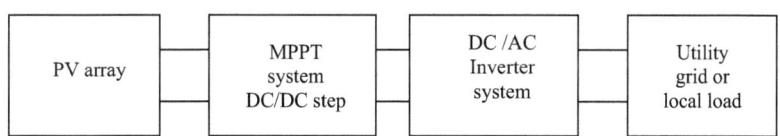

Figure 4.1: Block diagram of a solar PV grid connected application using MPPT

The PV arrays discussed in chapter 2 produce DC power. When a PV array is connected to utility grid, it is controlled by a MPP tracker. The DC power produced is converted to AC power through a DC-AC inverter. These functions are performed by the Power Conditioning System

(PCS), which is composed of the MPP tracker, inverter and its control system. The basic scheme of a solar PV grid connected system without any energy storage device is shown in Figure 4.2.

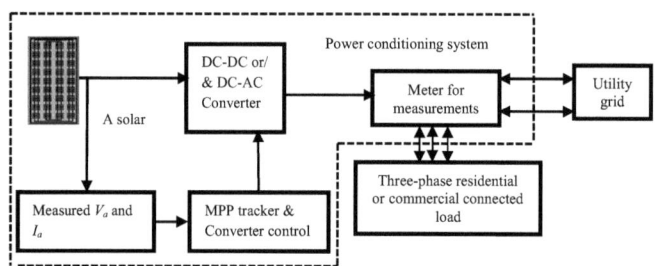

Figure 4.2: Typical connection scheme of a solar PV grid connected system

The PCS is a device, which interfaces the solar array with the utility grid. In a grid connected system without storage, the PCS consists of a MPPT technique and Voltage Source Converter (VSC) systems. The converter system is comprised of either a DC-AC (in single-stage) converter or DC-DC converter and DC-AC (in double-stage) converter and its controller. A MPPT technique is used for extracting the maximum power from the solar PV module and transferring that power to the load. A DC-DC converter (step-up or step-down) serves the purpose of transferring maximum power from the solar PV module to the load. By changing the duty cycle, the load impedance as seen by the source is varied and matched at the point of the peak power with the source, so that the maximum power is transferred. The PCS also contains protective devices, which disconnect the PV system from the grid, in the case of grid failure. The PCS may also use the existing controllable switches of the DC-AC inverter in order to perform these protective functions.

4.2.1 DC-AC Converter Systems

These systems are comprised of a power electronic interface, which inverts the DC power generated into a suitable AC current for the grid. It also controls the solar PV module, so as to track the MPP for maximizing the energy capture. This function is normally achieved by Pulse Width Modulation (PWM) at a fixed frequency, where the switching device is mainly Bipolar Junction Transistor (BJT), Metal Oxide Semiconductor Field Effect Transistor (MOSFET) or

Insulated Gate Bipolar Transistor (IGBT). The PWM control signal for the inverter is generated by comparing controlled voltage with a saw-tooth voltage [21]. The control system of inverter is comprised of two controllers. One controller is a MPPT controller, whereas the other one is controlling pulse for PWM generator. The PWM control is possible through different types of controlling techniques.

4.2.2. DC-DC Converter Systems

The DC-DC converters are used as switching mode regulators to convert an unregulated DC voltage to a regulated DC output voltage. There are four topologies for the switching regulators: buck converter, boost converter, buck-boost converter, and Cuk converter. The buck converter topology is used for voltage step-down, whereas the boost converter topology is used for stepping up the voltage. For PV system with batteries, the MPP of commercial PV module is set above the charging voltage of batteries for most combinations of solar radiation and temperature. Mostly, a buck converter is operated at MPP. However, it cannot operate at MPP when the point goes below the battery charging voltage under a low solar radiation and high temperature conditions. Thus, the additional boost capability slightly increases the overall efficiency. The PV grid connected systems uses a boost type converter to step up the output voltage to the utility level, before the inverter stage. The DC-DC boost converter is a step up DC voltage converter. When the boost converter operates in continuous conduction mode, the inductor current never falls to zero. The relation between input voltage V_i and the output voltage V_o is given as Equation 4.1,

$$\frac{V_o}{V_i} = \frac{1}{(1-D)}$$

(4.1)

From the above equation, it can be seen that the output voltage is always higher than the input voltage (as the duty cycle D varies from 0 to 1). It increases with D theoretically to infinity as D approaches 1. Due to this, this converter is sometimes referred to as a step-up converter.

An important aspect related to the PV system connected to the electric grid is that it can operate the double functions of active power generator and reactive power compensator. The proper power factor is selected according to active power and reactive power that the grid demands. At the same time, it can supply reactive power to the electrical grid when there is little or no solar radiation. That is important for compensating the reactive power at peak hours, when the main grid needs a amount of reactive power higher than average consumption. Although the photovoltaic system does not generate active power in such period of time, it can supply reactive power up to its maximum.

4.3 Power Voltage and Voltage Current Characteristics of a Typical PV Module

The output power [18] of a solar PV module changes with change in direction of sun, changes in solar radiation level and with varying temperature as shown in the Figure 4.3 and Figure 4.4. As seen in the *P-V* curve of the module there is a single maxima of power that is, there exists a peak power corresponding to a particular voltage and current. We know that the efficiency of the solar PV module is low about 13%. Since, the module efficiency is low it is desirable to operate the module at the peak power point so that the maximum power can be delivered to the load under varying temperature and radiation conditions. Hence, maximization of power improves the utilization of the solar PV module.

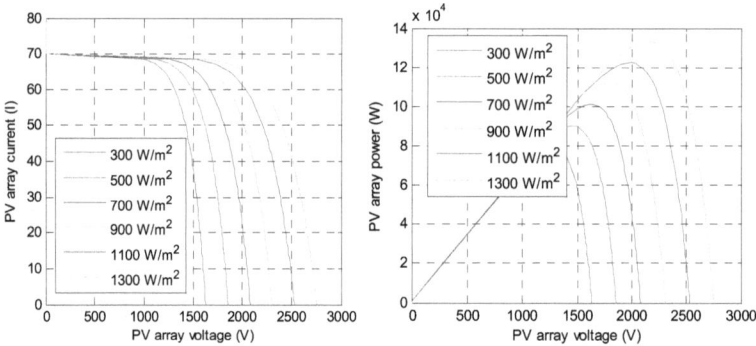

Figure 4.3: Changes in the V-I and P-V characteristics of the solar PV module due to change in radiation level.

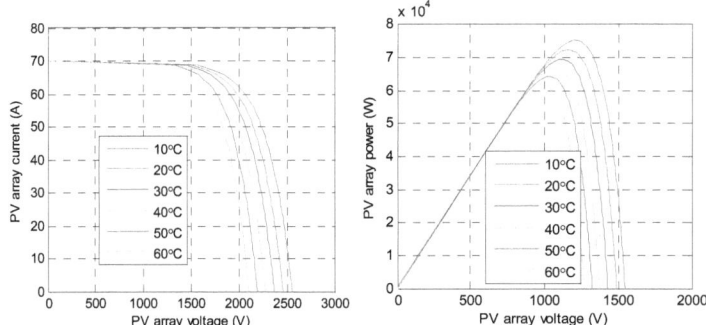

Figure 4.4: Changes in the V-I and P-V characteristics of the solar PV module due to change in temperature.

4.4 Classification of MPPT Techniques

There are many MPPT techniques available for extracting maximum power from solar PV modules and transfer to connected load. However, there are two types of basic MPPT techniques, which have widely been applied in solar PV grid connected systems:

(i) Perturb & Observe (P&O) technique, and
(ii) Incremental Conductance (IC) technique

4.4.1 P&O-MPPT Technique

Out of all basic MPPT techniques, P&O-MPPT is one of the efficient and commonly used techniques. This MPPT technique is very popular, because of its simplicity and ease of implementation. It is implemented in order that PV power can be generated efficiently even under changing weather conditions. The approach of the proposed technique is a simple, which calculates the direction in which to perturb the solar PV array's operating point to reach MPP. As shown by flow-chart in Figure 4.5, the principle of P&O controller is to provoke perturbation by increasing or decreasing the value of PWM duty cycle D and observe the resulting change in the output power. If the instantaneous PV output power $P_a(k)$ is greater than previously computed PV output power $P_a(k-1)$, then the same direction of duty cycle perturbation is maintained, otherwise the direction is reversed.

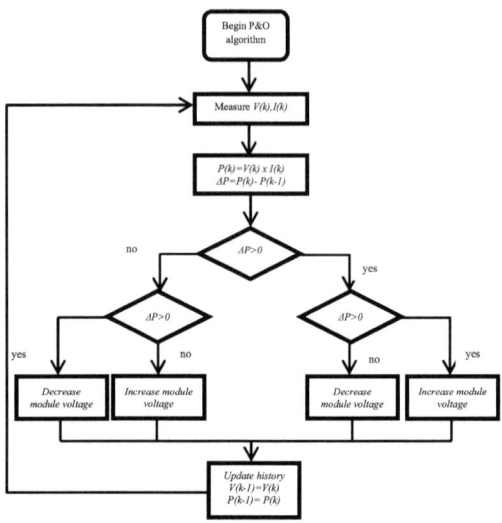

Figure 4.5: Flowchart of Perturb & Observe MPPT technique

From the P-V curve shown in Figure 4.6 (a) region I: If, $\frac{dP_a}{dV_a} > 0$ and $D(k) = D(k-1) + C$, there is increase in voltage; region II: if, $\frac{dP_a}{dV_a} < 0$ and $D(k) = D(k-1) - C$, then there is decrease in voltage, C is incrimination step. It is expected that using this MPPT technique, the maximum output power can be obtained, which can be controlled. Figure 4.6 (a) and Figure 4.6 (b) depicts the P-V and V-I curves obtained from simulations by using a 100 kW solar PV array (manufacturer datasheet specifications of PV model Sunpower SPR-305-WHT [19]). As shown, the red dots on blue curves indicate module manufacturer specifications (V_{oc}, I_{sc}, V_{mp}, I_{mp}) under standard test conditions (25°C, 1000 W/m²). These results show a good agreement for different weather conditions such as at 25°C, and 250/500/750/1000-W/m². This solar PV array exhibits highly non-linear P-V characteristics, which strongly depends on variable environmental conditions as shown in Figure 4.6 (c). Since MPP changes with variations in solar radiation and ambient temperature of a solar cell, the solar PV array has to be continuously operated within the region of MPP for an optimized utilization of the solar PV system.

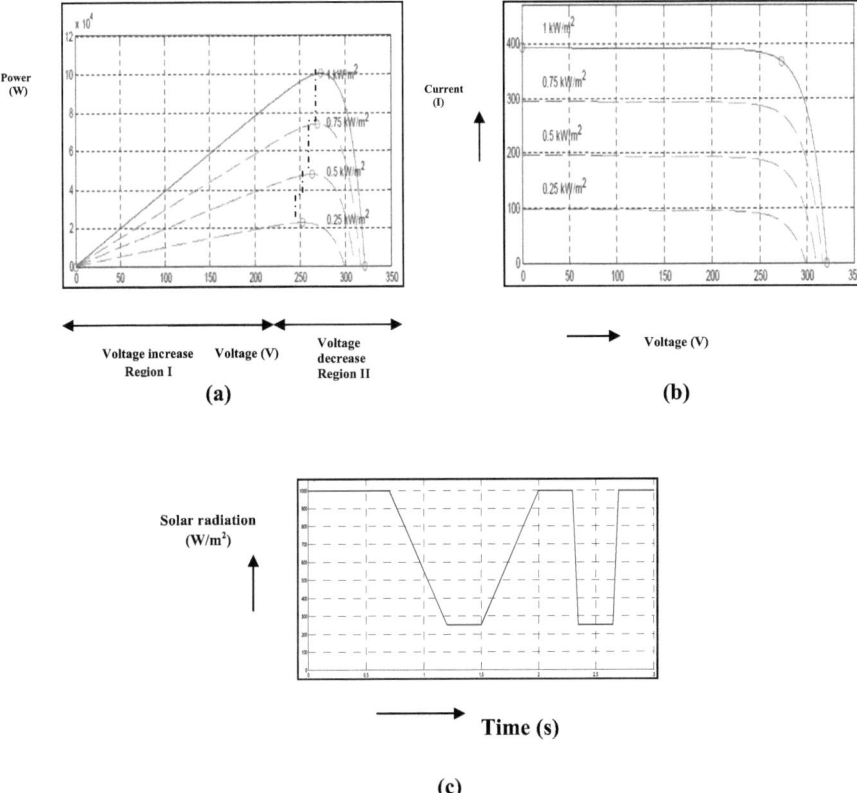

Figure 4.6 (a) Array P-V curve (b) Array V-I curve (c) Ramp up-down behavior of solar radiation intensity

From Figure 4.6, it can be seen that incrementing (decrementing) the voltage increases (decreases) the power when operating on the left of the MPP and decreases (increases) the power when on the right of the MPP. Therefore, if there is an increase in power, the subsequent perturbation should be kept the same to reach the MPP and if there is a decrease in power, the perturbation should be reversed, Table 4.1. The process is repeated periodically until the MPP is reached. The system then oscillates about the MPP. The oscillation can be minimized by reducing the perturbation step size. However, a smaller perturbation size slows down the MPPT. A solution to this conflicting situation is to have a variable perturbation size that gets smaller towards the MPP discussed in [20-22].

Table 4.1: Algorithm of P&O-MPPT Technique

Perturbation	Change in power	Next Perturbation
Positive	Positive	Positive
Positive	Negative	Negative
Negative	Positive	Negative
Negative	Negative	Positive

4.4.2 IC-MPPT Technique

Based on the various literature surveys, the traditional IC-MPPT technique is introduced briefly here. The principle of IC-MPPT algorithm flowchart shown by Figure 4.7, is derived by differentiating solar PV array power P_a (W) with respect to its voltage V_a (V), and equating result equal to zero as given by Equation 4.2,

$$\frac{dP_a}{dV_a} = \frac{dV_a I_a}{dV_a} = I_a + V_a \frac{dI_a}{dV_a} = 0 \text{ (at MPP)} \qquad (4.2)$$

Rearranging the above equation, which gives,

$$\frac{-I_a}{V_a} = \frac{dI_a}{dV_a} \qquad (4.3)$$

Hence left hand side of Equation 4.3 represents opposite of solar PV array's instantaneous conductance, while its right hand side represents the incremental conductance. These two quantities must be equal in magnitude, but opposite in sign at MPP. If operating point is off of MPP, a set of inequalities can be derived that indicates operating voltage is above or below MPP voltage.

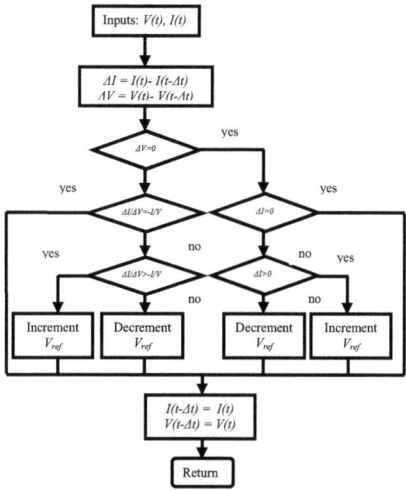

Figure 4.7: Flowchart of Incremental conductance MPPT technique

Efficiency Comparison: A comparison of the above mentioned two MPPT techniques has been done which is shown in Table 4.2

Table 4.2: Efficiency Comparison of P&O and IC-MPPT technique

Irradiance conditions (S, W/m²)	P&O-MPPT		IC-MPPT
	ΔV=2.5%	ΔV=5%	ΔV=5.0 %
500	99.9870	99.9559	99.9975
1,000	99.9885	99.9518	99.9875

The above comparison shows that IC-MPPT performs slightly better as compared to P&O-MPPT, therefore, this method is more suitable for the grid connected solar PV systems by using digital signal processors which reduces the computational effort of MPPT technique. However, the implementation of IC-MPPT algorithm is little more complex as compared to P&O-MPPT [23] technique.

4.5 Analysis of Other MPPT Techniques

A brief analysis of few other MPPT techniques is explained below. These techniques although are simple in nature, however, easy to implement and analyze.

4.5.1 Fractional Open-Circuit Voltage Technique

The near linear relationship between V_{MPP} and V_{OC} of the solar PV array, under varying irradiance and temperature levels, has given rise to the fractional open-circuit voltage method [24].

$$V_{MPP} = k\, V_{OC} \quad (4.4)$$

where k is a constant of proportionality. Since, k is dependent on the characteristics of the PV array being used, it usually has to be computed beforehand by empirically determining V_{MPP} and V_{OC} for the specific PV array at different irradiance and temperature levels. The factor k has been reported to be between 0.71 and 0.78. Once k is known, V_{MPP} can be computed using Equation 4.4 with V_{OC} measured periodically by momentarily shutting down the power converter. However, this incurs some disadvantages, including temporary loss of power. To prevent this, uses pilot cells from which V_{OC} can be obtained [25]. These pilot cells must be carefully chosen to closely represent the characteristics of the PV array. It is claimed that the voltage generated by pn-junction diodes is approximately 75% of V_{OC}. This eliminates the need for measuring V_{OC} and computing V_{MPP} [26].

4.5.2. Fractional Short-Circuit Current Technique

Fractional I_{SC} results from the fact that, under varying atmospheric conditions, I_{MPP} is approximately linearly related to the I_{SC} of the PV array,

$$I_{MPP} = k1\, I_{SC} \quad (4.5)$$

where $k1$ is proportionality constant. Just like in the fractional V_{OC} technique, $k1$ has to be determined according to the PV array in use. The constant $k1$ is generally found to be between 0.78 and 0.92. Measuring I_{SC} during operation is problematic. An additional switch usually has to be added to the power converter to periodically short the PV array so that I_{SC} can be measured using a current sensor. This increases the number of components and cost. A boost converter may be used, where the switch in the converter itself can be used to short the PV array [27].

4.6 Conclusion

This chapter has highlighted the needs of PV installation components other than PV panels. These components are jointly referred to as the Balance of System (BOS) and include the batteries, DC-DC converters, DC-AC converters for AC loads and grid connected systems. Since *P-V* and *V-I* curves of a solar panel are non-linear in nature, the need and importance of MPPT technique in extracting variable solar power at maximum points is described in PCS systems.

CHAPTER V: INTRODUCTION TO POWER QUALITY

5.1 Introduction

Today, the electric power organizations are no longer being operated as independent ones. They are being part of big network of utility companies, which are connected together in a complex network of grid. Usually, it is required that the heavy electrical machines and switchgear devices in big industry require the electric power which is distortion-less. This quality of electric power shows that the good Power Quality (PQ) is essential for smooth functioning of interconnected electric grid network. This is due to the fact that majority of loads in distribution system are inductive in nature. As per Institution of Electrical and Electronic Engineering (IEEE)-1100, PQ is defined as *"the concept of powering and grounding sensitive electronic equipment in a manner that is suitable to the operation of that equipment."* The various parameters which affect PQ are found in reference [28] viz. variations in voltage magnitude, harmonics in AC power waveform, transient and oscillatory nature of current-voltage, with continuity of service.

During these days, all interconnected electric power systems are complex in nature, where majority of power generating stations and load centers are interconnected. Here, the major concerns for customers are reliability of continuity and quality of power supply available at load centers. It is observed that the electric power generation in most of developed countries is reliable, but quality of power supply is not.

The power supply system can only control the quality of the voltage. It has no control over the currents that particular loads might draw. Therefore, the standards in the PQ area are devoted to maintaining the supply voltage within certain limits. Any significant deviation in the waveform magnitude, frequency or purity is a potential PQ problem. Of course, there is always a close relationship between voltage and current in any practical power system. Although the generators may provide a near-perfect sine-wave voltage, the current passing through the impedance of the system can cause a variety of disturbances to the voltage. PQ is often considered as a combination of voltage and current quality. In most of the cases, it is considered that the network operator is responsible for voltage quality at the point of connection while the customers load often influences the current quality at the point of connection.

5.2 Sources of Poor Power Quality

The main sources of poor PQ in any electric power system are listed below:
- Adjustable speed drives
- Switching power supplies
- Arc furnaces
- Electronic fluorescent lamp ballasts
- A lightning strike
- Non- Linear loads
- Starting of large induction motors
- Power electronic devices

5.3 Need of Power Quality Concerns

Since most of the industrial loads are non-linear in nature, there is an increased concern of PQ due to the following reasons:
- New-generation loads that use microprocessor and microcontroller based controls devices, are more sensitive to PQ variations than that equipment's used in the past.
- The demand for improved system efficiency provided growth of some devices like speed motor drives etc. for correcting the power factor and also for reducing losses. This results in increased level of harmonics in power systems.
- The client users have an awareness of issues related to PQ. Clients are now becoming knowledgeable about these issues like sag and swell etc. [29].

5.4 Power Quality Problems

The various PQ problems, which have been found from the literature survey are enumerated below:

- **Voltage sag:** It is also referred to as a voltage dip, Figure 5.1 and is the most common PQ problem in electric power system. It is defined as the decrease of Root Mean Square (rms) voltage from its maximum value to a value between 0.1 p.u. to 0.9 p.u. It lasts for duration between 0.5 cycles to 1 minute.

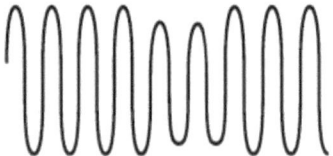

Figure 5.1: Waveform for voltage sag

- **Voltage swell:** A voltage swell is defined as an increase in rms voltage from its maximum value to in between 1.1 and 1.8 p.u. at power frequency during 0.5 cycles to 1 minute. As shown in Figure 5.2, a voltage swell likewise sag, is also characterized by its rms magnitude and duration. The voltage swell is mainly caused by switching-off large capacitors, start and stop of heavy inductive loads.

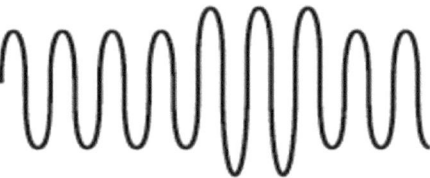

Figure 5.2: Waveform for voltage swell

- **Voltage interruption:** As shown in Figure 5.3, a voltage interruption is defined a the decrease in rms system voltage to less than a small percentage of the nominal rated voltage, which is also called a complete loss of rms voltage. Voltage interruption may arise from faults, component malfunctions, and scheduled downtime. As compared, the short voltage interruptions are typically the result of malfunction of switching device or a deliberate or inadvertent operation of any switchgear, or its re-closure in response to any fault and system disturbance.

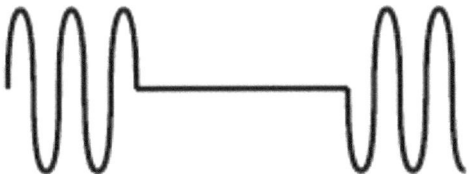

Figure 5.3: Waveform for voltage interruption

- **Spikes:** A spike is a sudden or short surge in system rms voltage. The voltage spike is caused by lightning, short circuits, power transitions, or power outages in large equipment on the same power line.
- **Transients:** The transients also known as surge, are PQ disturbances that involve high magnitudes of destructive current and voltage or both. In magnitude, it may attain thousands of volts and amperes, even in low voltage type of systems. This type of PQ phenomenon exists in a very short duration from less than 50 nano-seconds to as long as 50 milli-seconds. The various sources of transients are: switching activities, lightning strikes, capacitor bank switching, re-closing operations, tap changing on transformers, loose connections in any distribution system that results arcing, accidents, opening and closing of disconnects of energized lines, human error, animals and bad weather conditions and neighboring facilities.
- **Impulsive transient:** It is a type of transient disturbance which enters into any electric power system. This PQ issue is defined according to IEEE-1159 standard, as a sudden, non-power frequency change in the steady-state condition of rms voltage, current, or both which is unidirectional in polarity-either primarily positive or negative. For example, Lightning. The currents resulted from any lightning strike can reach as high to several thousand amps in about 2-3 μs time duration.
- **Oscillatory transients:** The oscillatory transient is defined as a sudden, non-power frequency change in steady-state condition of voltage, current, or both that has both positive and negative polarity values (bidirectional).
- **Voltage fluctuations and flickers:** The voltage fluctuations are systematic variations of the rms voltage envelope as a series of random changes in the voltage magnitude. It lies in the range of 0.9 p.u. to 1.1 p.u. It is mainly caused by high power loads that draw

fluctuating current, such as large motor drives and arc furnaces. It causes low frequency cyclic voltage variations, which results in flickering of light sources such as incandescent and fluorescent lamps. Thereby it can cause significant irritation in human beings.

- **Waveform distortion:** It is defined as a steady-state deviation from an ideal sine wave of power frequency. Different types of waveform distortion are as follows:

a) Harmonics: A harmonic is any sinusoidal frequency, which is a multiple of the fundamental frequency. Harmonic frequencies can be even or odd multiples of the sinusoidal fundamental frequency. The main causes for harmonic distortion are rectifiers and all non-linear loads, such as power electronics equipment.

b) Notching: Notching is a periodic voltage disturbance caused by the normal operation of power electronic devices when current is commutated from one phase to another, shown in Figure 5.4.

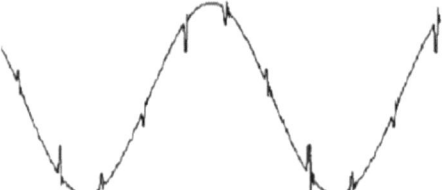

Figure 5.4: Waveform for notching

c) Noise: Noise is defined as unwanted electrical signal with broadband spectral content lower than 200 kHz superimpose upon the power system voltage or current in phase conductors, or found on neutral conductors or signal lines, Figure 5.5.

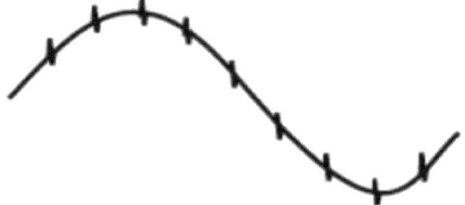

Figure 5.5: Waveform for noise

5.5 Solutions Adopted to Improve Power Quality

The solution to improve PQ can be done from customer side and utility side. The various approaches, which are used to improve it are as follows:

- **Load conditioning:** It ensures that operated equipment is less sensitive to power frequency disturbances, thus, allowing operation even under significant voltage distortion.
- **Line conditioning systems:** This system suppresses or counteracts the various power system disturbances. To improve PQ, passive filters are connected at the sensitive load terminals. However, the main challenge is to regulate sensitive load terminal voltage in order that its magnitude remains constant and any distortion caused by harmonic is reduced to an acceptable level.

5.6 Power Quality Standards

Standards are needed to achieve coordination between the characteristics of the power supply system and the requirements of the end use equipment. This is the role of PQ standards. The methods have been established for measuring these phenomena and in some cases defining limits for satisfactory performance of both the power system and connected equipment. In the international community, both IEEE and International Electro-technical Commission (IEC) have created a group of standards that addresses these issues from a variety of perspectives [30].

5.7 Standards Related with Voltage Characteristics

The most common international standards setting limits on voltage quality are described below.

5.7.1. IEEE Standards

Short duration voltage variations are the variations that occur in the fundamental frequency voltage with less than one minute. These variations are best characterized by plots of the rms voltage versus time but it is often sufficient to describe them by a voltage magnitude. Voltage variations can be a voltage sag, voltage swell or loss of voltage. IEEE-1159 standard specifies durations for instantaneous, momentary and temporary disturbances. The most common index used is system average rms frequency index. This index represents the average number of voltage sags experienced by an end user each year with a specified characteristic.

5.7.2. IEC Electromagnetic Compatibility Standards

An inclusive structure of standards on Electromagnetic Compatibility (EMC) is under expansion within the IEC. EMC is defined as: the ability of a device, equipment or system to function satisfactorily in its electromagnetic environment without introducing intolerable electromagnetic disturbances to anything in that environment. There are two aspects to EMC: A piece of equipment should be able to operate normally in its environment and it should not pollute the environment too much. In EMC terms: immunity and emission. There are standards for both aspects. Immunity standards define the minimum level of electromagnetic disturbance that a piece of equipment shall be able to withstand.

The basic immunity standard IEC-61000-4-1 gives four classes of equipment performance as follows:

(i) Normal performance within the specification limits

(ii) Temporary degradation or loss of function which is self-recoverable

(iii) Temporary degradation or loss of function which requires operator intervention or system reset

(iv) Degradation or loss of function which is not recoverable due to damage of equipment, components or software or loss of data.

5.7.3. The European Voltage Characteristics Standards

European Nation (EN)-50160 dealing with requirements concerning the supplier's side characterizes voltage parameters of electrical energy in public distribution systems. On the user's side, it is the quality of power available to the user's equipment that is important. Correct equipment operation requires the level of electromagnetic influence on equipment to be maintained below certain limits. Equipment is influenced by disturbances on the supply and by other equipment in the installation, as well as itself influencing the supply. These problems are summarized in the EN-61000 series of EMC standards, in which limits of conducted disturbances are characterized. European- 50160 standard gives the main characteristics of the voltage at the customer's supply terminals in public low-voltage and medium-voltage networks under normal operating conditions.

5.8 Standards Related With Current Characteristics

The most common international standards setting limits on harmonics are described in the following subsections:

5.8.1. IEEE Standards

For the current harmonic limits, Total Demand Distortion (TDD) calculation is used as given in Equation 5.1. The term TDD is very much like Total Harmonic Distortion (THD) as given in Equation 5.2. The only difference is the dominator. THD calculation compares the momentary measured harmonics with the momentary measured fundamental component. TDD calculation compares the momentary (but steady-state) measured harmonics with the maximum demand current, which is not a momentary number at all. The difference between TDD and THD is important because it prevents a user from being unfairly penalized for harmonics during periods of light load (only the harmonic polluting loads are running). During periods of light load it can appear that harmonic levels have increased in terms of percent (THD calculation) even though the actual current harmonics in amperes (TDD calculation) stayed the same.

$$\text{TDD} = \frac{\text{RMS harmonic Current}}{\text{Max. Demand Load Current}} \quad (5.1)$$

$$\text{THD} = \frac{\text{RMS Sum of all Harmonic Currents}}{\text{RMS Fundamental Currents}} \quad (5.2)$$

Individual Harmonic Distortion (IHD) at a particular harmonic frequency is the ratio of the rms of the harmonic under consideration to the rms value of the fundamental as stated in Equation 5.3.

$$\text{IHD} = \frac{\text{Harmonic Frequency}}{\text{Fundamental Frequency}} * 100 \quad (5.3)$$

As the problems caused by harmonics become recognized around the world, standards setting bodies are creating electrical standards that define legal limits for the level of current harmonics and voltages.

5.8.2. The International Electro Technical Commission

IEC has a standard, IEC-61000-3-2, that defines current harmonic limits for devices with a current rating less than or equal to 16A. This has been ratified as a harmonized European Standard, EN 61000-3-2 and as a British standard. Unlike its predecessor (IEC 555-2, 1982), no distinction is made between domestic and professional equipment, rack mounted and three-phase equipment is specifically mentioned in BS EN 61000-3-2.

5.9 Conclusion

In this chapter, the introduction of PQ and its concerns has been presented. The various PQ issues have been introduced and discussed according to IEEE and IEC standards. The effect of each of PQ issue has been described for a distribution system according to IEC standard. It is discussed that in electric power systems, hundreds of power generating stations and load centers are interconnected and operated. There are various PQ problems likewise, voltage sag, voltage swell, transients, voltage interruption, harmonics, noise and notching. Each PQ problem can be solved by the proper coordination among all power system components.

CHAPTER VI: MATLAB & SIMULINK SOFTWARE

6.1 Introduction

Previous chapters of this book have presented the discussion on solar PV technology systems and various PQ concerns in any distribution system. This chapter presents the introduction to Matlab and simulink software in order that the study for any solar PV grid connected system can be demonstrated and analyzed conveniently. Matlab software helps any power engineering and researchers carrying PQ study using various tools available likewise power system tools, power electronics tools and application of controllers. Therefore, simulation for any designed system can be carried out demonstrating the impact of solar PV systems into existing electric power system.

6.2 Introduction to Matlab Software

MATLAB stands for **MAT**rix **LAB**oratory which deals in operation on matrices. The basic data type is an array, all data type i.e. vectors, scalars are treated as matrices. There is no need to declare the dimension of a matrix or type of variable. This software is developed by 'The Mathworks Incorporated' USA. This software package is used for numerical, scientific and engineering computations and visualization.

Matlab provides Integrated Development Environment (IDE) for Built in functions (beta function, gamma function, data analysis, laplace transform, fourier transform, polynomials and ordinary differential functions), user defined functions, external interface to run Fortran/ C/C++ features availability and collection of optional tools for specific applications.

Why MATLAB ??

C and other languages are difficult because it involves functions like declaring variables, specifying data types, and allocating memory. Matlab functions are built upon LAPACK (Linear Algebra PACKage written in Fortran-90) and BLAS (Basic Linear Algebra Subprograms). Combined with leading-edge methods, these highly optimized routines give access to the fastest, most robust numerical routines available.

Matlab provides interactive environment, hundreds of in-built functions for computations, 2-D and 3-D graphics, animation, and several toolboxes which are collections of functions written for special applications. All are easy to use shallow learning curve i.e. more learning with less effort. High speed of computations supports almost every computational platform like windows, UNIX, LINUX, Solaris operating systems.

6.3 Matlab Windows

There are six windows Matlab windows which are discussed below:
- Command window
- Command history window
- Workspace window
- Current directory window
- Graphics/Figure window
- Edit window

Command window: It is a main window, which has a Matlab command prompt ">>". All Matlab Commands are typed in this window and are typed after the prompt symbol ">>".

Command history window: Go to the command history window and double click on any of the commands listed. This selected command will appear in the command window. Select all the commands in the command history window by clicking on the date and right click the mouse. By right clicking, following options are made available:

Cut, copy, evaluate selection, create M-file, delete and selection.

Select create M-file from the options. Editor window will be opened and will have all these commands typed in that window. Thus, various commands can be tested and can be turned into M-files easily

Workspace window: Click on the tab for this sub-window. Select any of the variables t, x or y listed in this sub-window and double click to open an array editor window. Any value in the array can be changed here.

Graphics/Figure window: The output of all graphics/plot commands, typed in the command window is shown/flashed in the graphics/figure window. Depending on the system memory, many figure windows can be created by the user.

Edit window: In this window, user can write, edit and save his/her own programs in files called, M-files.

6.4 Matlab File Types

MATLAB has three types of files for storing information:
- M-files, which are standard ASCII text file
- Mat-files, which are binary data files (cannot be read by the user directly)
- Mex-files, which are callable FORTRAN/C program file

6.5 Simulink Software

Simulation link is an extension of Matlab for modeling, simulating, and analyzing dynamic, linear/nonlinear, complex control systems. Graphical User Interface (GUI) and visual representation of simulation process by simulation block diagrams are two key features which make Simulink one of the most successful software packages, particularly suitable for control system design and analysis.

Simulation block diagrams are nothing but the same block diagrams we are using to describe control system structures and signal flow graphs. Simulink offers a large variety of ready-to-use building blocks to build the mathematical models and system structures in terms of block diagrams. Block parameters should be supplied by the user. Once the system structure is defined, some additional simulation parameters must also be set to govern how the numerical computation will be carried out and how the output data will be displayed.

To start Simulink, enter **simulink** command at the Matlab prompt. Alternatively, one can also click on Simulink icon shown in Figure 6.1.

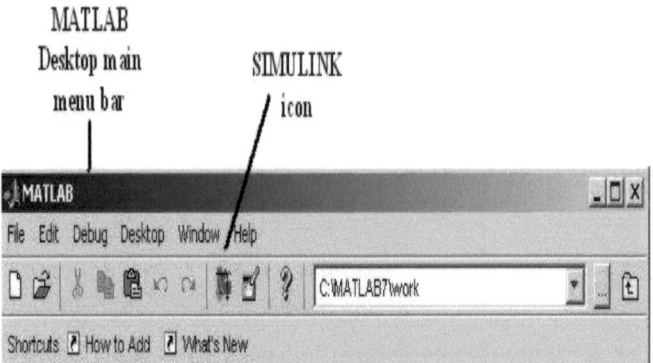

Figure 6.1: Matlab desktop main menu and Simulink icon

A Simulink library browser as shown in Figure 6.2 appears which displays tree-structured view of the Simulink block libraries. It contains several nodes; each of these nodes represents a library of subsystem blocks that is used to construct simulation block diagrams. You can expand/collapse the tree by clicking on the ⊞/⊟ boxes beside each node and block in the block set pan.

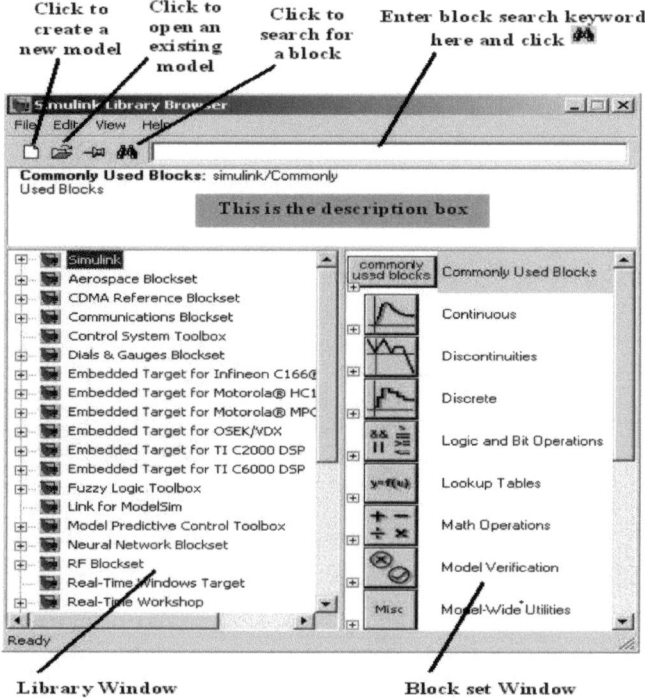

Figure 6.2: Simulink library browser

From the above window, user can expand the node labeled Simulink. Thereafter, sub-nodes of this node (Commonly Used Blocks, Continuous, Discontinuities, Discrete, Logic and Bit Operations) are displayed. Now for example, expanding the sources sub-node displays a long list of sources library blocks. User can simply click on any block to learn about its functionality in the description box (see Figure 6.3).

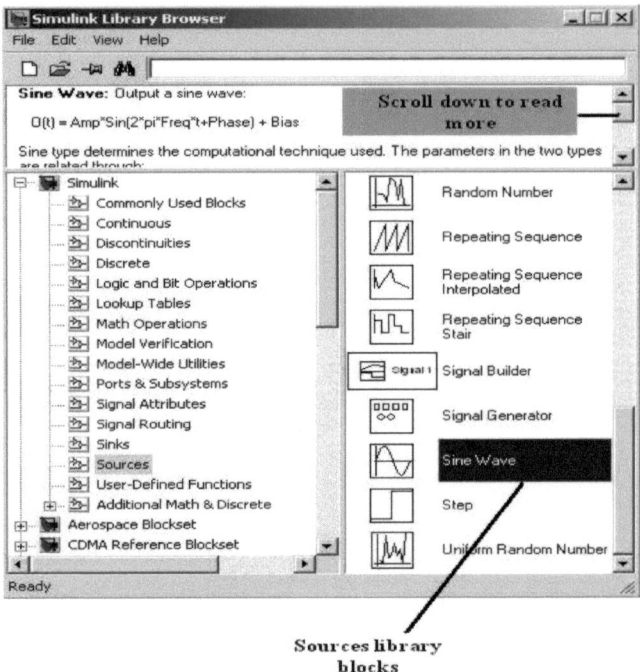

Figure 6.3: Blocks in Sources sub-node

The sources sub-node may now be collapsed and expand the sinks sub-node. A list of sinks library block appears, Figure 6.4. Learn the purpose of various blocks in sinks sub-node by clicking on the blocks.

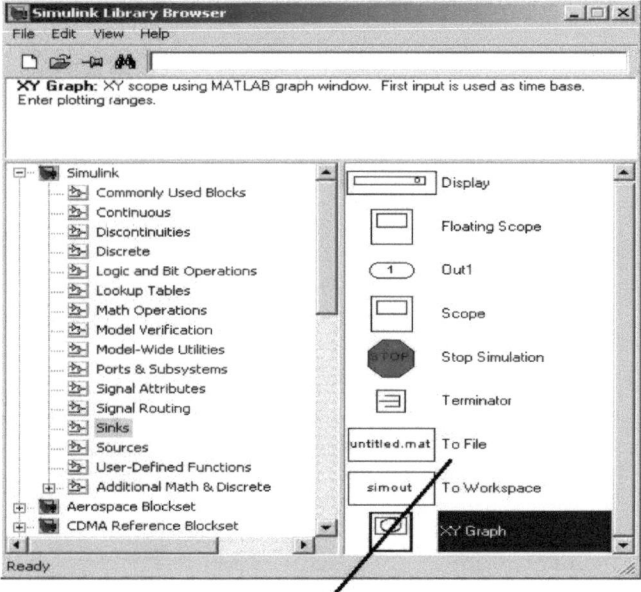

Figure 6.4: Blocks in Sinks sub-node

Identification of block(s) required for simulation, is the first step of the construction of simulation diagram in Simulink. The next step is to "drag and drop" the required blocks from Simulink block libraries to untitled workspace, Figure 6.5. Right clicking on the block will provide various options to users from which one can cut, copy, delete, format (submenu provides facilities for rotation of the block, flipping, changing the font of block name).

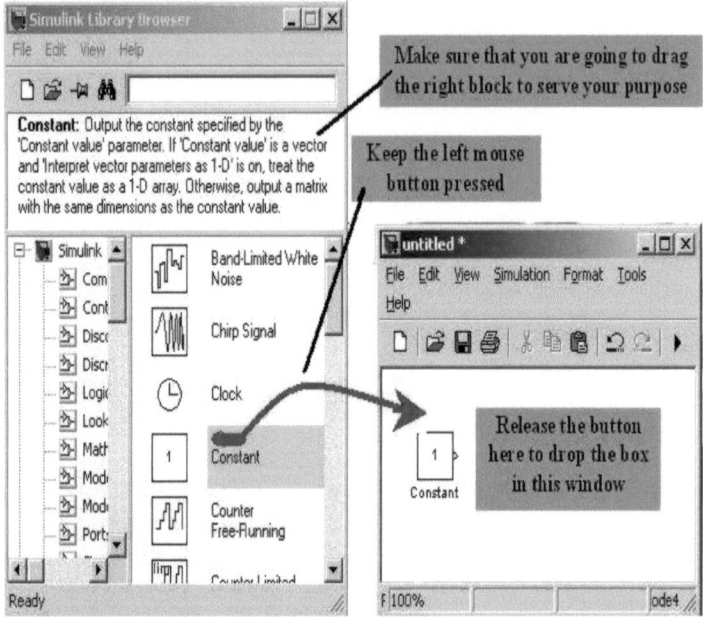

Figure 6.5: Drag and drop blocks to workspace from library browser

Copy the rest of the blocks in a similar manner from their respective libraries into the model window. User can move a block from one place in the model window to another by dragging the block. User can move a block a short distance by selecting the block, then pressing the arrow keys. With all the blocks copied into the model window, the model should look something as shown in Figure 6.6.

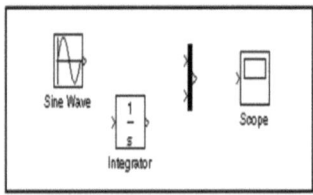

Figure 6.6: Components for a Simulink model

On examining the blocks, user see an angle bracket on the right of the Sine Wave block and two on the left of the Mux block. The > symbol pointing out of a block is an output port; if the symbol points to a block, it is an input port. A signal travels out of an output port and into an input port of another block through a connecting line. When the blocks are connected, the port symbols disappear. Now it's time to connect the blocks. User can connect the Sine Wave block to the top input port of the Mux block. Position the pointer over the output port on the right side of the Sine Wave block. User can notice, that the shape of cursor changes to cross hairs.

User should hold down the mouse button and move the cursor to the top input port of the Mux block. On releasing the mouse button, the blocks will be connected. User can also connect the line to the block by releasing the mouse button while the pointer is over the block. On doing so, the line is connected to the input port closest to the cursor's position. However, one line connects a line to the input port of another block. This line, called a branch line, connects the Sine Wave output to the Integrator block, and carries the same signal that passes from the Sine Wave block to the Mux block. To complete a connection to an existing line, following steps are followed:

- Position the pointer on the line between the Sine Wave and the Mux block.
- Press and hold down the Ctrl key (or click the right mouse button). Press the mouse button, then drag the pointer to the Integrator block's input port or over the Integrator block itself.
- Release the mouse button. Simulink draws a line between the starting point and the Integrator block's input port.

Finish making block connections, user can see the model should look as shown in Figure 6.7.

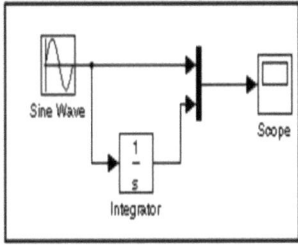

Figure 6.7: Complete model of Simulink connected

71

6.5.1 Controlling Execution of a Simulation

The Simulink graphical interface includes menu commands and toolbar buttons that enable user to start, stop and pause a simulation.

6.5.2 Starting a Simulation

To start execution of a model, user can select Start from the model editor's Simulation menu or click the Start button on the model's toolbar, shown in Figure 6.8.

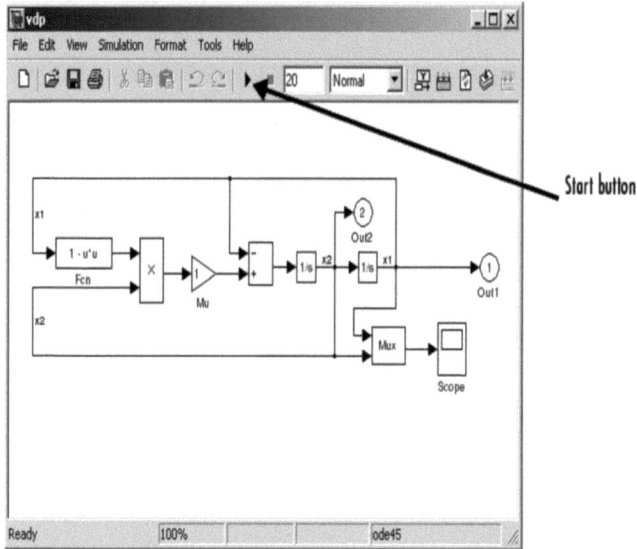

Figure 6.8: To simulate a Simulink model

User can also use the keyboard shortcut, **Ctrl+T,** to start the simulation.

Note: A common mistake that new Simulink users make is to start a simulation while the Simulink block library is the active window. Make sure, user model window is the active window before starting a simulation. Simulink starts executing the model at the start time specified on the Configuration Parameters dialog box. Execution continues until the simulation reaches the final

72

time step specified on the Configuration Parameters dialog box, an error occurs, or user can pause or terminate the simulation.

While the simulation is running, a progress bar at the bottom of the model window shows how far the simulation has progressed. A **Stop** command replaces the **Start** command on the Simulation menu. A **Pause** command appears on the menu and replaces the **Start** button on the model toolbar. The computer beeps to signal the completion of the simulation as shown in Figure 6.9.

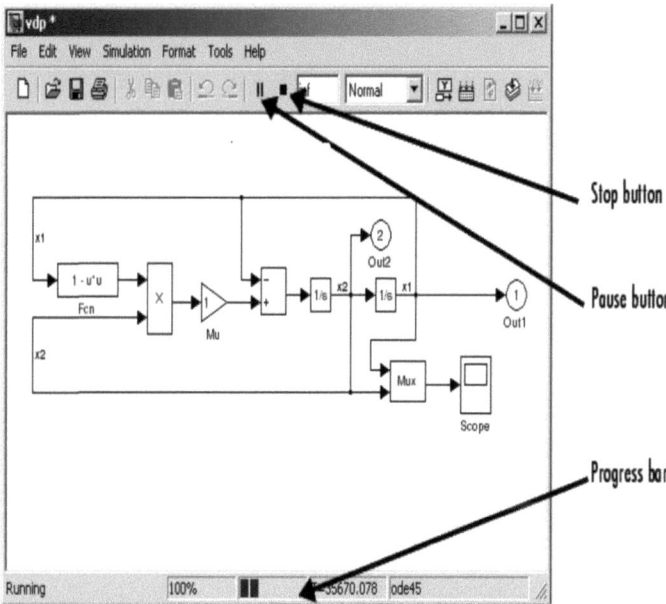

Figure 6.9: Options of Simulink simulation

6.5.3 Ending a Simulink Session

Terminate a Simulink session by closing all Simulink windows. Terminate a Matlab session by choosing one the **File** menu and **Exit Matlab**.

6.6 Conclusion

In this chapter, it has been shown that Matlab is a software package, which can be used to perform analysis and solve mathematical and engineering problems. Introduction to various

Matlab windows have been given and described. Simulink contains a library editor of tools from which we can build input/output devices and continuous and discrete time model simulations. Simulink has a comprehensive block library which can be used to simulate linear, non-linear or discrete systems. C codes can also be generated from Simulink models for embedded applications and rapid prototyping of control systems. This chapter also highlights the procedure for the operation of Simulink software with building blocks.

CHAPTER VII: SOLAR PV GRID CONNECTED SYSTEM WITH MPPT CONTROL

7.1 Introduction

In previous chapter of this book, the introduction on Matlab software is explained with the opening of Simulink interfacing. The basic procedure of Matlab-Simulink has been demonstrated using various simulation windows. In this chapter, the simulink modeling of solar PV single- stage transformer-less grid connected system is introduced. The solar PV array tested in reference [14] has been chosen which can work even during the changing environmental conditions. The introduction of the solar PV array has been explained using basic semiconductor equations in chapter 2. In this chapter, the detail model including the VSC control, and the data based MPPT grid interactive control has been explained. The performance of the proposed system has been evaluated at linear *RLC* load.

7.2 System Computation Model

As shown in Figure 7.1, the solar PV array has been connected to VSC through the capacitor. The voltage across this capacitor has to maintain constant during changing conditions. The PCS is comprised of VSC along with its control, MPPT control; Phase Locked Loop (PLL) and *LCL* filter for the removal of harmonics. PLL algorithm is used for grid synchronization. The output of the VSC is connected to grid through an inductance so that the flow of power can take place on both sides. The presence of inductance creates a phase angle which is responsible for the flow of active and reactive from both sides.

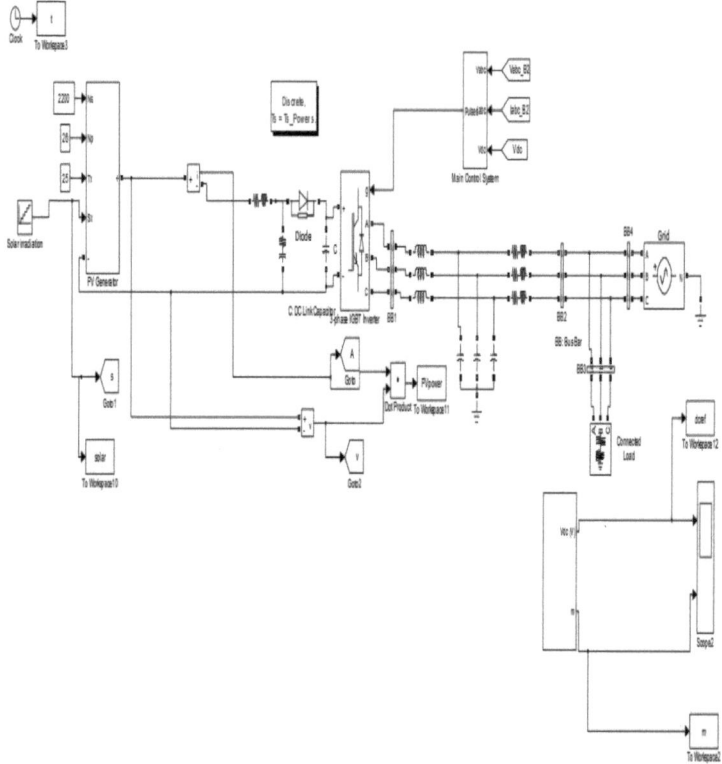

Figure 7.1 Simulink model of solar PV grid connected system

7.2.1. Rotating Reference Frame Transformation

For a three-phase sinusoidal signal, the following equations are obtained after *abc_to_dq0* transformation [31]:

$$V_d = \frac{2}{3}(V_a \sin(\omega t) + V_b \sin(\omega t - \frac{2\pi}{3}) + V_c \sin(\omega t + \frac{2\pi}{3})) \tag{7.1}$$

$$V_q = \frac{2}{3}(V_a \cos(\omega t) + V_b \cos(\omega t - \frac{2\pi}{3}) + V_c \cos(\omega t + \frac{2\pi}{3})) \tag{7.2}$$

$$V_o = \frac{2}{3}(V_a + V_b + V_c) \quad (7.3)$$

where V_a, V_b and V_c is the signal from the three-phase VSC voltage, whereas V_d, V_q and V_o are the direct-axis, quadrature-axis and zero-sequence component of voltage.

7.2.2 Real & Reactive Power Control

Typically, the efficiency of IGBT-VSC is very high. In the absence of active input power, the grid connected converter is operated in reactive power generation mode, which powers the control circuitry, compensates the converter losses, and maintains a regulated DC voltage. With simulation results, it has been shown that, when real power is not available, the DC link capacitor is charged and voltage is kept within limits, while injecting the desired level of reactive power into the grid. The diagram in Figure 7.2 shows two power sources coupled with an impedance of Z, whose $\phi = \tan^{-1}\frac{X}{R}$

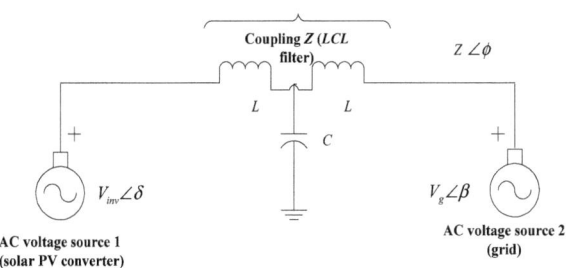

Figure 7.2: Power flow between the two AC sources

Source 1 and source 2 represents the solar PV converter system and the utility grid, respectively. Source 1 is identified with $V_{inv}\angle\delta$ and source 2 with $V_g\angle\beta$ where V represents the rms voltage and angle represents the phase reference. If power flows from source 1 to source 2 through the coupling Z [32], the current flow I can be defined as:

$$I = \frac{(V_{inv}\angle\delta - V_g\angle\beta)}{R + jX} = \frac{(V_{inv}\angle\delta - V_g\angle\beta)}{Z\angle\phi} \quad (7.4)$$

Using $S = VI^*$, we get:

$$S = V_{inv}\angle\delta \frac{(V_{inv}\angle\delta - V_g\angle\beta)^*}{Z\angle\phi} \tag{7.5}$$

$$S = \frac{(V_{inv}V_{inv}\angle(\delta-\delta) - V_{inv}V_g\angle(\delta-\beta))}{Z\angle-\phi} \tag{7.6}$$

$$S = \frac{V_{inv}^2\angle\phi}{Z} - \frac{V_{inv}V_g\angle(\delta-\beta)}{Z} \tag{7.7}$$

$$S = \frac{V_{inv}^2}{Z}\cos\phi + j\frac{V_{inv}^2}{Z}\sin\phi - \frac{V_{inv}V_g\cos(\delta-\beta+\phi)}{Z} - j\frac{V_{inv}V_g\sin(\delta-\beta+\phi)}{Z} \tag{7.8}$$

$$S = \frac{V_{inv}^2}{Z}\cos\phi - \frac{V_{inv}V_g\cos(\delta-\beta+\phi)}{Z} + j(\frac{V_{inv}^2}{Z}\sin\phi - \frac{V_{inv}V_g\sin(\delta-\beta+\phi)}{Z}) \tag{7.9}$$

$$Re[S] = P = \frac{V_{inv}^2}{Z}\cos\phi - \frac{V_{inv}V_g\cos(\delta-\beta+\phi)}{Z} \tag{7.10}$$

$$Im[S] = Q = \frac{V_{inv}^2}{Z}\sin\phi - \frac{V_{inv}V_g\sin(\delta-\beta+\phi)}{Z} \tag{7.11}$$

Real power flow: Now, considering Equation 7.10, if the phase reference β for source 2 (grid) is taken as the reference or unity power factor operation at grid ($\beta = 0$), we get

$$P = \frac{V_{inv}^2}{Z}\cos\phi - \frac{V_{inv}V_g\cos(\delta+\phi)}{Z} \tag{7.12}$$

Assuming R is very small for only inductive coupling, $R \cong 0, \phi \cong 90°, Z = jX$, Equation 7.12 becomes,

$$P = \frac{V_{inv}^2}{Z}\cos 90° - \frac{V_{inv}V_g\cos(\delta+90°)}{Z} \tag{7.13}$$

$$P = \frac{V_{inv}V_g\sin\delta}{Z} \tag{7.14}$$

Reactive power flow: Considering Equation 7.11, and assuming $\beta = 0, R \cong 0, \phi \cong 90°, Z = jX$, we get

$$Q = \frac{V_{inv}^2}{X} - \frac{V_{inv}V_g \sin(\delta + 90°)}{X}$$

$$Q = \frac{V_{inv}^2}{X} - \frac{V_{inv}V_g \cos\delta}{X}$$

$$Q = \frac{V_{inv}}{X}(V_{inv} - V_g \cos\delta) \qquad (7.15)$$

$V_g \angle 0$ represents the grid of $230 \angle 0$ (230 V rms and 0° phase reference) and X is of fixed value. From the above analysis of power transfer theory, it is found that when R is close to 0 Ω, the loss across R is very small, and the power transfer depends as follows:

(i) Real power mainly depends on δ.
(ii) Reactive power mainly depends on V_{inv} (rms value of source 1).

The converter current is given as, $I_{inv} = \dfrac{\sqrt{2}\sqrt{P^2 + Q^2}}{V_g}$

$$(7.16)$$

From Equation 7.14 to Equation 7.15, it is clear that the real power mainly depends on δ whereas, the reactive power depends on V_{inv}. A positive value of real power P implies feeding real power into the grid, while a negative one results in drawing power from the grid.

7.2.3. Synchronization and Control of Three-Phase Grid Connected Inverter System

One of the most important and necessary features of a power converter connected to electric utility grid is proper synchronization with the three-phase voltages in a three-phase system. The synchronization methods used for three-phase systems are more complex than in single-phase systems due to the relationship of phase shift and phase sequence of the coordinated three-phase voltages. The modulus and the rotational speed of the three-phase voltage vector are maintained constant when balanced sinusoidal waveforms are present in the three-phase system. The balanced system of voltages is represented as,

$$V_{abc} = \begin{bmatrix} V_a \\ V_b \\ V_c \end{bmatrix} = V \begin{bmatrix} \cos(\omega t + \phi) \\ \cos(\omega t - \frac{2\pi}{3} + \phi) \\ \cos(\omega t + \frac{2\pi}{3} + \phi) \end{bmatrix}$$

(7.17)

The three-phase power converters used with PV arrays injects positive-sequence current at fundamental frequency into the grid. These converters inject negative-sequence and harmonic currents in abnormal cases, depending on the purpose of the converter. Therefore, grid synchronization of at three-phase system requires an advanced detection system designed to reject both higher order harmonics, which detects the sequence components in a quick and accurate manner. PLL's are employed in order to track the angular frequency and phase shift of the three-phase voltages or more precisely positive-sequence components of the three-phase voltages, for synchronization. Various advanced PLL techniques have been proposed in literature, but a simple and easy to implement software based; three-phase discrete PLL is used on the extracted positive- sequence of the three-phase voltages. The PLL is capable of synchronizing the inverter operation to the utility grid, in order to test the proposed ideas.

7.2.4. Generation of PWM Pulses

PWM is generated using sine triangle PWM waveform. In order to produce the output voltage of desired magnitude waveform, phase shift and frequency, the desired signal is compared with a carrier (triangular waveform signal) of higher frequency to generate appropriate switching signals, as depicted in Figure 7.3. The DC link capacitor is alternately connected to the inverter outputs with positive and negative polarity. When the switches are closed at on time, the voltage time averaging over one carrier wave begins. The control of on-time and off-time is achieved by comparing the modulating voltage with the carrier voltage. When the magnitude of the carrier voltage exceeds the magnitude of the modulating voltage, one of the active switches is opened to end any contribution to the time average voltage. Similar triangles on the control plot of voltage vs. time show that,

$$\frac{T}{T_s} = \frac{V_{carrier} - V_{modulation}}{V_{carrier}}$$

(7.18)

$$V_{average} = \frac{T_s - T}{T_s} X \frac{V_{dc}}{2} = \frac{V_{modulation}}{V_{carrier}} X \frac{V_{dc}}{2} = m \frac{V_{dc}}{2} \qquad (7.19)$$

where the modulation index, *m* varies with time to synthesize the average voltage. If the average voltage were plotted, it would look like the modulating voltage waveform (inverter sine output). The output voltage of the VSC does not have the shape of the desired signal, but switching harmonics can be filtered out by the series *LCL* low pass filter. It retrieves the 50 Hz fundamental sine wave.

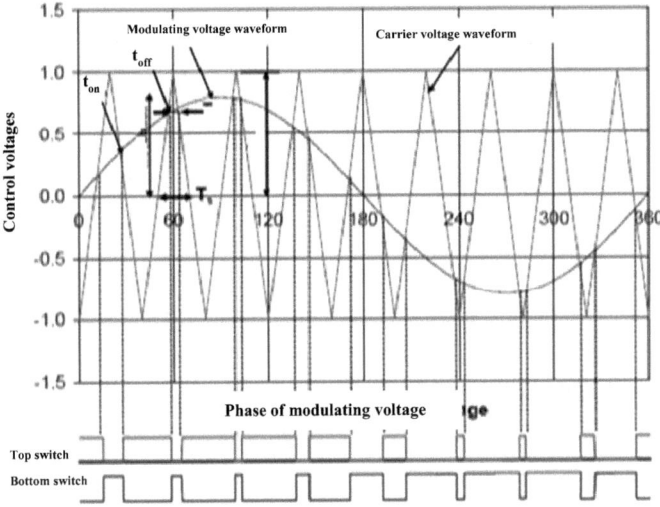

Figure 7.3: Generation of PWM pulses

SPWM is one of the most popular modulation techniques [33] among others applied in power switching inverters. In this technique, the output voltage (line to line) as obtained in linear modulation range is given by,

$$V_{AB} = V_{BC} = V_{CA} = mV_{dc} \frac{\sqrt{3}}{2} \qquad (7.20)$$

$m \leq 1.0$, for linear range. In SPWM switching technique, the magnitude and phase angle of voltages at inverter output directly depend on modulation index, and its phase angle. The inverter's voltages (V_A, V_B, V_C) are expressed in terms of the upper switches as follows:

$$\begin{pmatrix} V_a \\ V_b \\ V_c \end{pmatrix} = \frac{V_{pv}}{3} \begin{pmatrix} -2 & -1 & -1 \\ -1 & 2 & -1 \\ -1 & -1 & 2 \end{pmatrix} \begin{pmatrix} K_1 \\ K_2 \\ K_3 \end{pmatrix} \quad (7.21)$$

where K_1, K_2 and K_3 are the controller signals applied to the switches.

7.3 Controlling Scheme of Voltage Source Converter

Figure 7.4 depicts the controlling scheme of VSC where the PLL and synchronous reference frame algorithms are used. Synchronous reference frame transforms the three phase signal into direct axis and quadrature axis components which can be analyzed easily. SPWM is one of the most widely used modulation techniques which is applied in switching of power inverters. In this technique, the line to line output voltage in linear modulation range is given by:

$$V_{AB} = V_{BC} = V_{CA} = m \frac{\sqrt[2]{3}}{2} V_{dc} \quad (7.22)$$

The actual value of DC link voltage to IGBT based inverter is compared with reference voltage of MPPT control. In this work, the data based MPPT control is implemented the operation of which is integrated with VSC control. This voltage is compared in the PI type voltage controller. Then the proportional error is passed through PI controller. It generates reference value of direct axis signal current. Figure 7.4 depicts PI type current controller. Here, the current signal is converted into the voltage signal. The quadrature component of the current is kept zero for operating the system at unity power factor.

The modulation index m and angle δ are calculated as, respectively,

$$\text{Modulation index, m} = \sqrt{V_d^2 + V_q^2} \quad (7.23)$$

$$\text{Angle, } \delta = \tan^{-1}(\frac{V_d}{V_q}) \quad (7.24)$$

Using Equation 7.23 in Equation 7.22, it is found that DC-AC converter output voltage can be changed by changing the modulation index. When output voltage of the VSC is higher than grid voltage, the reactive power is supplied by developed solar PV system to grid. The reactive power is absorbed, when the inverse action takes place. Figure 7.5 shows the simulink blocks through which the real and reactive power calculations are completed.

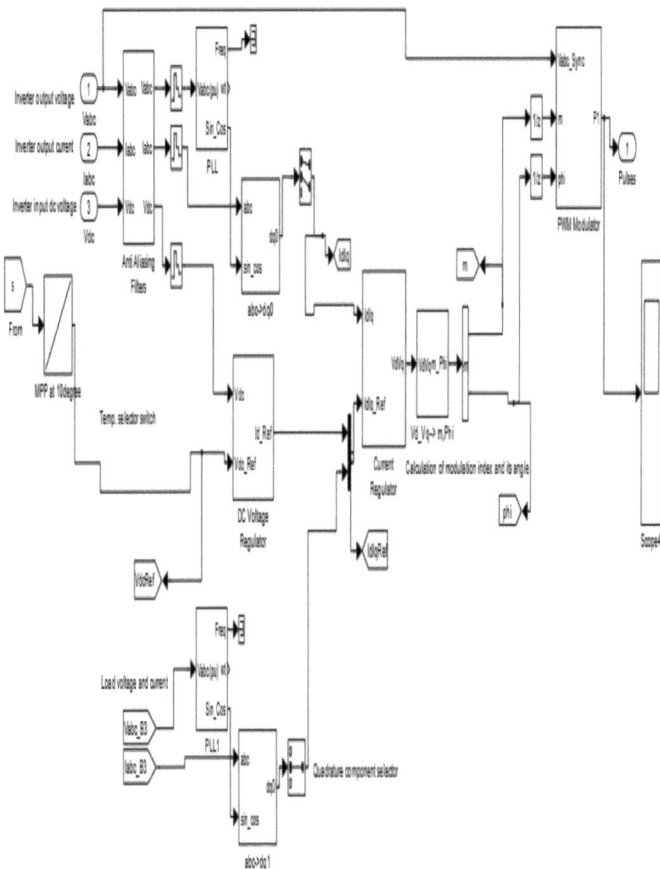

Figure 7.4: Simulink control model of voltage source converter

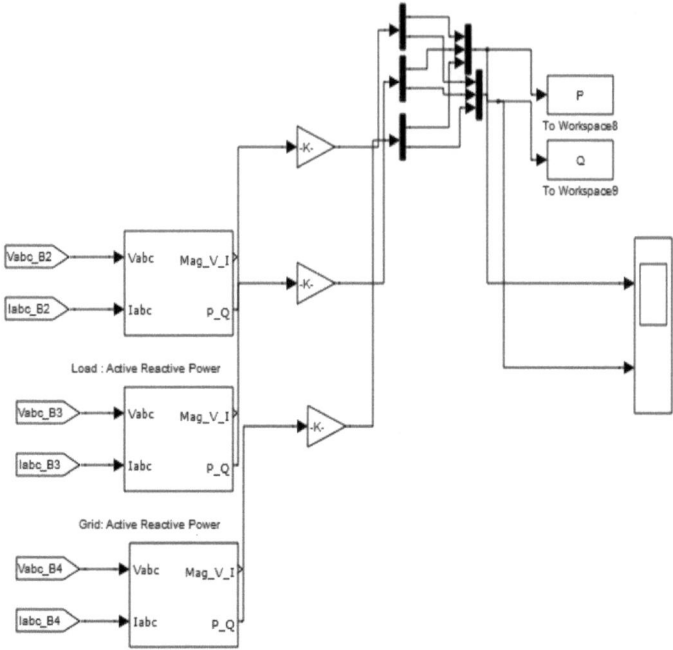

Figure 7.5: Simulink active and reactive power measurements

7.4 Data MPPT Control for Maximum Power Point

In this indirect type MPPT technique, the data is generated using seven solar PV modules forming an array. Each solar PV module is simulated at under changing atmospheric conditions i.e. changing solar radiation levels and ambient temperature at the specific cell temperature T_c. The data thus generated at maximum power is used in Matlab simulation using look up table as tabulated in Table 7.1. The output power of a single PV module depends upon output voltage, changing temperature and level of solar radiation. The MPPs are used for a particular solar radiation level at a cell temperature. Table 7.1, shows the simulated data through look-up table (with its solving method Interpolation-extrapolation in Matlab simulation).

Table 7.1: Data used at solar radiation S_x and temperatures T_x values

Vector of input values look up table	Solar radiation S_x (W/m^2)	Temperature T_x (^{0}C)	Maximum Power Point PV Voltage in volts at different cell temperature (Volts) $T_c = 10^0$C
0	88	10	600
20	90	20	700
40	92	30	800
60	94	40	900
80	96	50	1000
100	98	60	1100
120	100	70	1200

Table 7.2: System configuration parameters

System Parameters	Value chosen	System Parameters	Value chosen
Solar PV module - Boltzmann's constant - Photo-current -Reverse saturation current of diode - Series resistance of cell - No. of cells in series - No. of cells in parallel -Identity factor -Electron charge	1.38×10^{-23} J K^{-1} 6 A 0.0002 A 0.00011 Ω 2200 26 1.65 1.602×10^{-19} C	Anti-aliasing filters Second order low pass	Cut-off frequency 2 kHz
Capacitor: DC link	1500 µF (each)	Discrete 3-Ø PLL -sample time	f=50 Hz, f_{min} = 45 Hz 50×10^{-6} s
3-Ø IGBT VSC - Snubber resistance - Snubber capacitance	1×10^5 Ω Infinite	DC voltage regulator - Proportional gain, K_p - Integral gain, K_i	0.45 18
LCL filter	L=1500 × 10^{-6} H C=30 µF	Current Regulators - Proportional gain, K_p - Integral gain, K_i	0.025 5
3-Ø series RLC load	440 V,50 Hz, P=63 kW, Q=20 kW (+ve)	Discrete 3-Ø PWM generator (2-level)	f_c = 5 kHz
Utility grid at UPF	440 V(+ve sequence), 50 Hz		

7.5 Simulation Results and Discussion

The proposed system configuration parameters are listed in Table 7.2. The simulation time period is 0.3 s. Figure 7.6 (a) shows the nature of the solar radiation which acts as input to the proposed system. Since array current is directly proportional to the solar radiation, its nature is approximately proportional to the solar radiation curve, Figure 7.6 (b). Figure 7.6 (c) gives the PV array voltage whereas Figure 7.6 (d) shows the PV array power. The output power of PV array is product of PV array voltage and current. It is possible to observe that the approximate value of the PV array power is 48 kW.

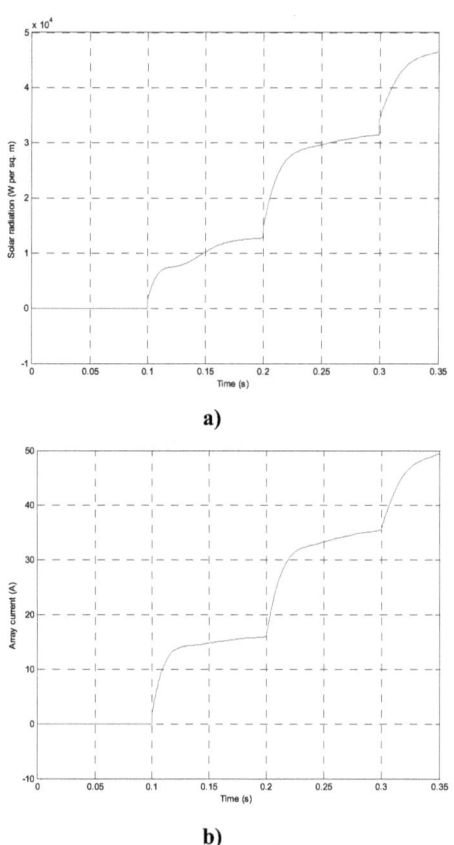

a)

b)

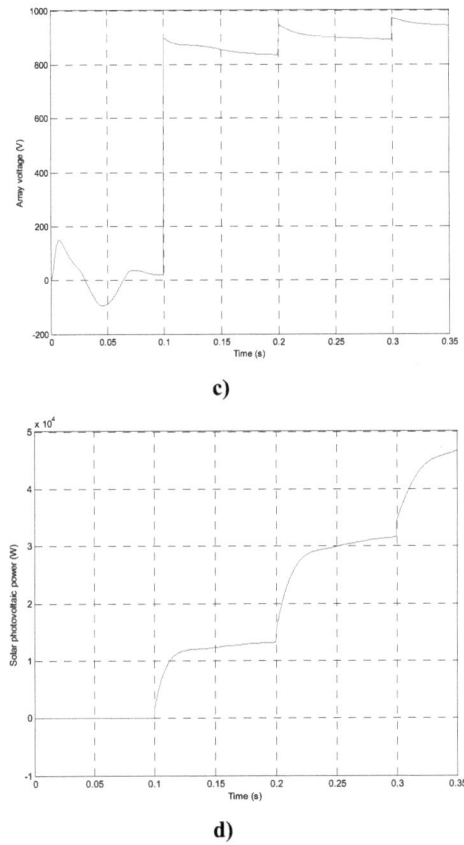

Figure 7.6: (a) Solar radiation in W/m2 (b) Solar array current (c) Solar array voltage (d) Solar array power

Figure 7.7 shows the nature of actual DC link voltage around reference MPPT voltage. The reference voltage has been generated from Table 7.1. The blue line indicates tracking of actual DC voltage around reference MPPT voltage. As the ambient temperature T_x increases and solar radiation increases, the actual value closely follows the reference voltage. The initial transients can be controlled by the further tuning of PI controllers. Figure 7.8 (a) shows that as the solar radiation level increases the grid power decreases. The power requirement of the load is met by PV array through VSC converter. This has been shown by Figure 7.8 (b) which depicts the VSC power. Figure 7.9 shows the variation of the modulation index.

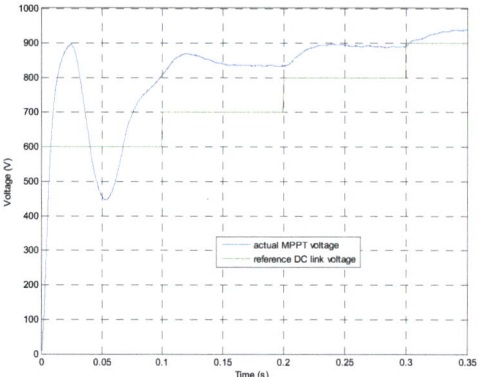

Figure 7.7: actual DC link voltage and MPPT reference voltage

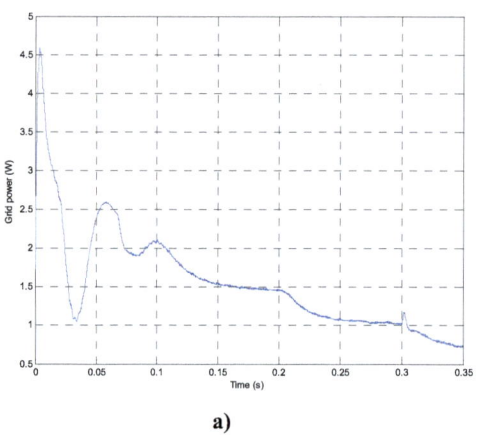

a)

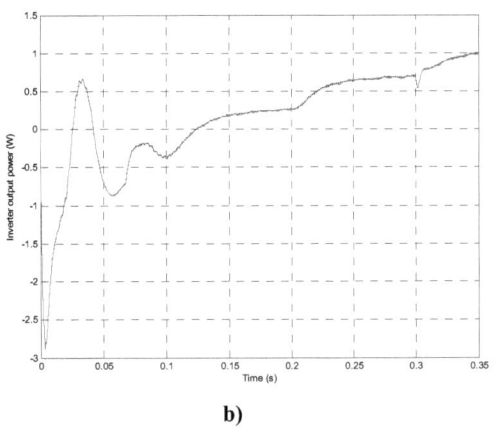

b)

Figure 7.8: (a) Grid output power (b) Inverter output power

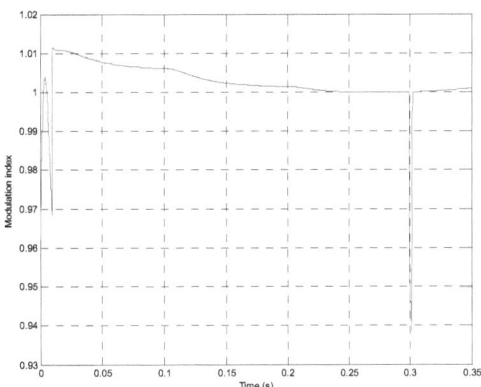

Figure 7.9: Modulation index

7.6 Conclusion

This chapter has presented the modeling of solar PV single stage grid connected system at unity power factor. No transformer is used in the proposed system as it increases the level of harmonics in the overall system. The nature of real power generated by solar PV array through VSC has been shown and proved that whenever the power from grid is un-available, the real power requirement of the load is achieved by VSC. Data based MPPT is proposed through which behavior of actual DC link voltage is discussed.

CHAPTER VIII: EVALUATION AT DIFFERENT POWER FACTOR & FREQUENCY

8.1 Introduction

In this chapter of book, the effect of changing power factor and frequency grid side on the proposed system discussed in previous chapter is discussed. The response of proposed system at unity power factor has been presented in chapter 7. Here, active and reactive power variation has been discussed under changing variable conditions. In addition, THD study has been carried out to study the level of power quality when any distributed generation source likewise solar is integrated with conventional grid connected systems.

8.2 Simulation Results and Discussion

In order to validate the proposed system, the performance of the system is evaluated at different power factors and frequencies. The total simulation period has been chosen as $t = 0.35$ s. The following are the two case studies:

Case I at different power factors and same frequency

Case II at different frequencies and same power factors

8.2.1 Case I at different power factors and same frequency

- *at unity power factor, 440 V, 50 Hz*

As shown in Figure 8.1 (a), the active power is generated at different temperature and solar radiation, at unity power factor. The system has been tested at constant frequency of 50 Hz. This active power has been generated according to data generated through data-based MPPT. The grid power completes the load power demand. As it decreases, the load power demand is met with the solar PV array through the IGBT based converter system. The initial transient at $t = 0.02$ s can be more controlled by introducing tuning of PI controllers.

Similarly, the reactive power response has been shown in Figure 8.1 (b). Reactive power compensation of the load is done by solar PV through converter system (shown by blue line). At $t = 0.05$ s, the reactive power generated by grid decreases till $t = 0.35$ s. The reactive power of the load is shown as constant at 20 kVAR.

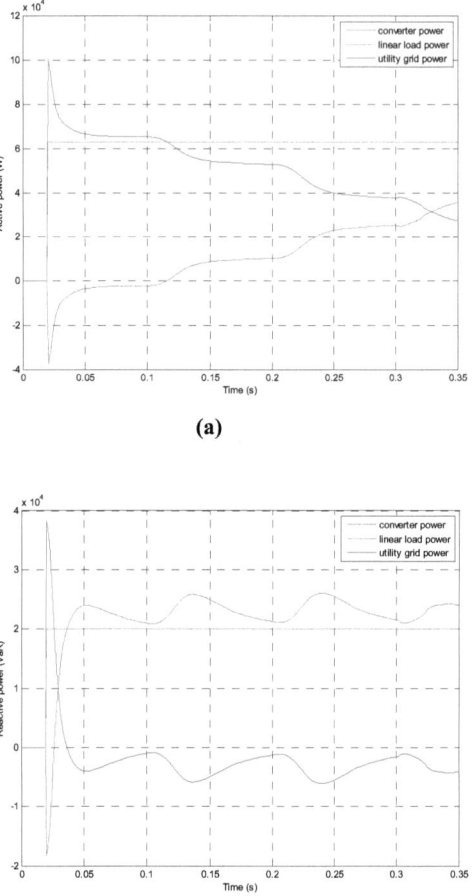

Figure 8.1: (a) Active power (b) Reactive power, variation of VSC, load and utility grid

Figure 8.2 (a) and Figure 8.2 (b) shows the THD response of VSC current and grid injected current at unity power factor, respectively. THD values from Fast Fourier Transform (FFT) analysis is found to be 1.05 % and 0.49 %, respectively.

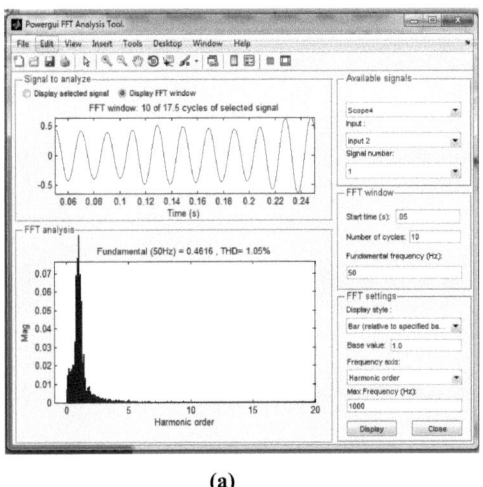

(a)

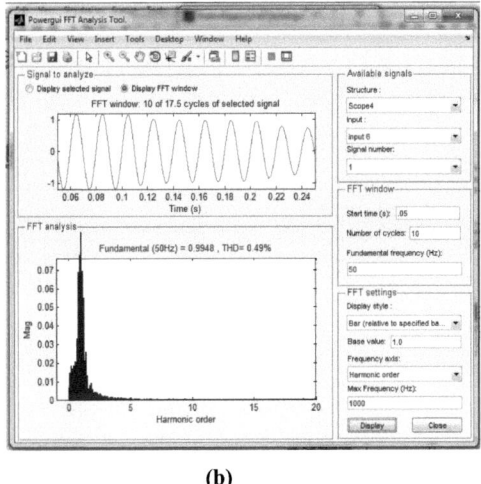

(b)

Figure 8.2: FFT analysis of (a) VSC current and (b) grid current (Unity power factor, 50 Hz)

- *at power factor = 0.866, 440 V, 50 Hz*

As depicted by Figure 8.3 (a), there has been a non-uniformity in the active power generation between the period $t = 0.07$ s to $t = 0.1$ s, which is due to decrease in the power factor from unity to 0.866. This is due to the fact, that the power factor is directly proportional to active power. Similar behavior has been shown by reactive power response between the period $t = 0.02$ s to $t = 0.1$ s, Figure 8.3 (b).

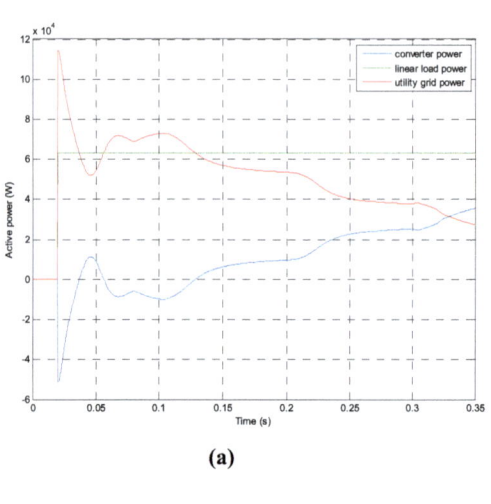

(a)

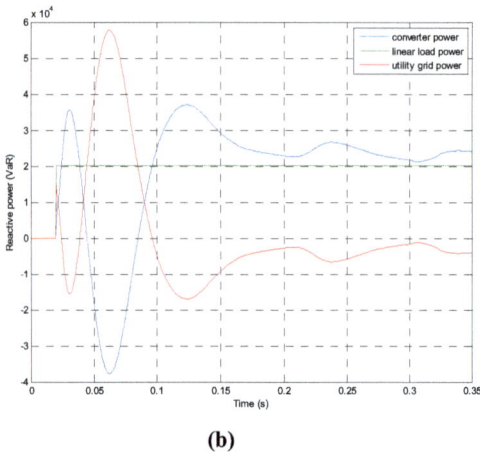

(b)

Figure 8.3: (a) Active power (b) Reactive power, variation of VSC, load and utility grid

Figure 8.4 (a) and Figure 8.4 (b) shows the THD response of VSC current and grid injected current at 0.866 power factor, respectively. The THD values are found to be 7.19 % and 2.71 %, respectively. Thus there has been the increase in THD with decrease in power factor.

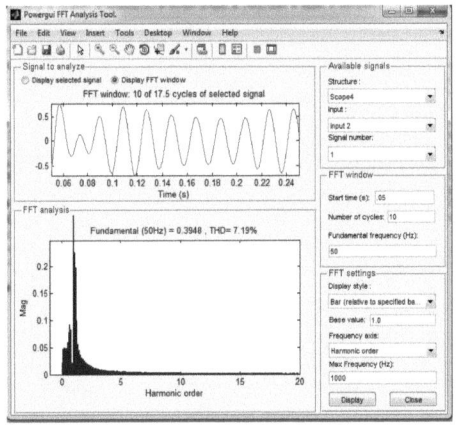

(a)

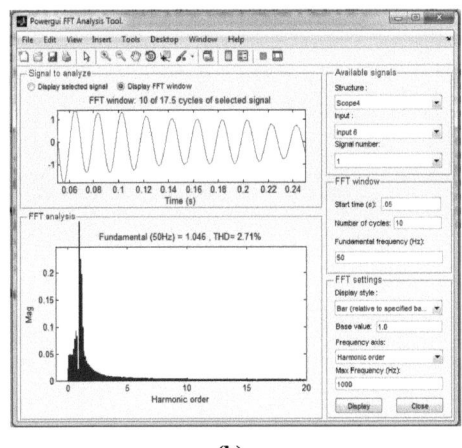

(b)

Figure 8.4: FFT analysis of (a) VSC current and (b) grid current

- *at power factor = 0.707, 440 V, 50 Hz*

At 0.707 power factor, the response becomes more oscillatory as shown by active power generated by solar PV through converter system, Figure 8.5 (a). The maximum power generated by utility grid is 80 KW. However, there has been decrease in the overshoots in the reactive power response after $t = 0.15$ s, Figure 8.5 (b).

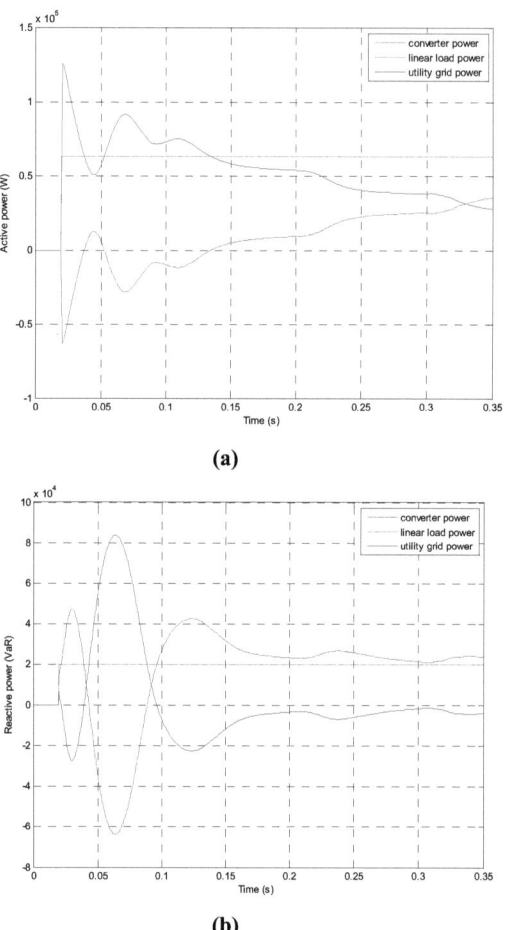

Figure 8.5: (a) Active power (b) Reactive power, variation of VSC, load and utility grid (Power factor=0.707, 50 Hz)

Figure 8.6 (a) and Figure 8.6 (b) shows the THD response of VSC current and grid injected current at 0.707 power factor, respectively. The THD values are found to be 8.28 % and 2.67 %, respectively. Again, there has been the increase in the THD of VSC current with decrease in power factor.

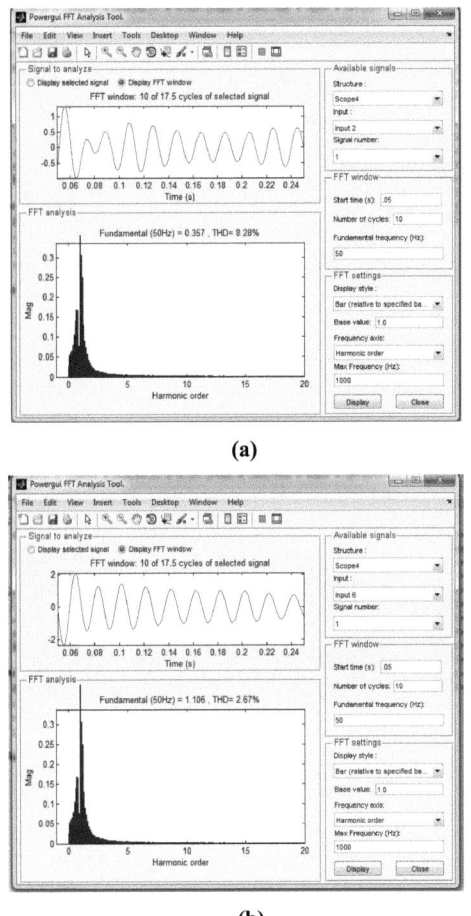

Figure 8.6: FFT analysis of (a) VSC current and (b) grid current (Power factor=0.707, 50 Hz)

- *at power factor = 0.5, 440 V, 50 Hz*

Figure 8.7 (a) depicts that the overshoot increases to 1.2×10^5. Thus the oscillatory response increases with the decrease in operating power factor. From Figure 8.7 (b), the reactive power generated by utility grid is 1.1×10^5. However, in this case, the compensation is met with the grid power from 0.04 s to 0.08 s, thereafter, completed by reactive power generated by VSC.

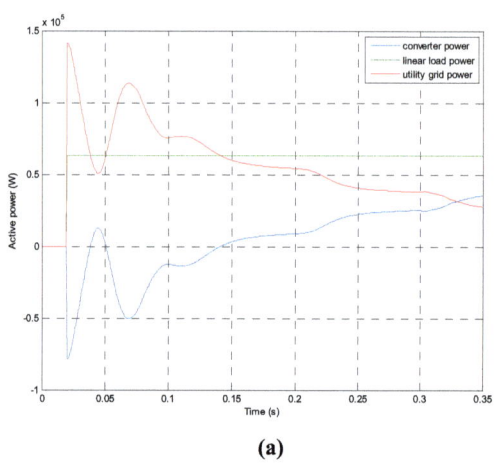

(a)

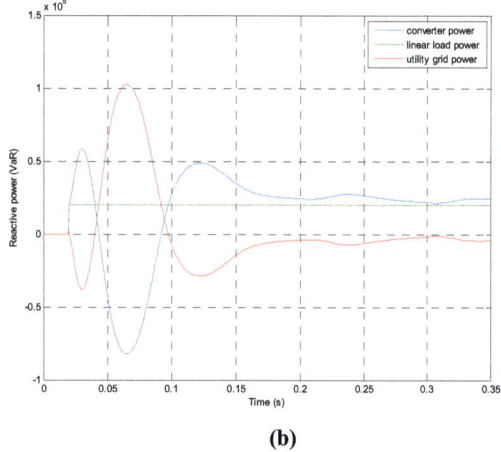

(b)

Figure 8.7: (a) Active power (b) Reactive power, variation of VSC, load and utility grid (Power factor=0.50, 50 Hz)

Figure 8.8 (a) and Figure 8.8 (b) shows the THD response of VSC current and grid injected current at 0.707 power factor, respectively. The THD values are found to be 9.68% and 2.58%, respectively. Thus, it has been found that there has been the increase in the THD of VSC current from 1.05% to 9.68% as the power factor decreases.

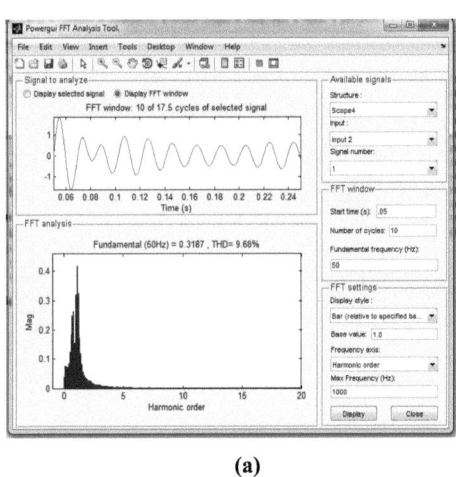

(a)

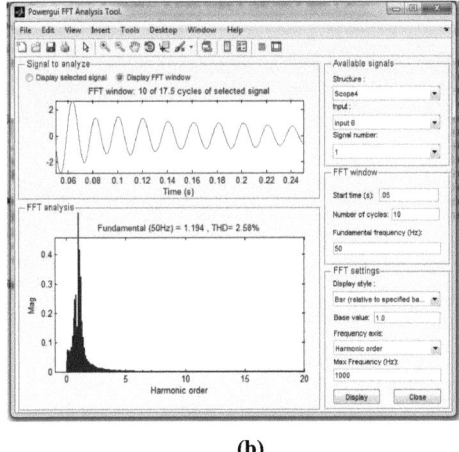

(b)

Figure 8.8: FFT analysis of (a) VSC current and (b) grid current (Power factor=0.50, 50 Hz)

8.2.2. Case II at different frequency and same power factors

This sub-section presents the behavior of proposed system at unity power factor and two different frequencies.

- *at unity power factor, 440 V, 49 Hz*

Figure 8.9 (a) shows real output power response whereas; Figure 8.9 (b) shows reactive output power response of VSC, connected load and electric grid. Figure 8.10 (a) and Figure 8.10 (b) shows the THD response of VSC current and grid injected current at unity power factor, respectively. However, this response has been obtained at 49 Hz frequency. The THD values are found to be 0.49% and 1.84%, respectively.

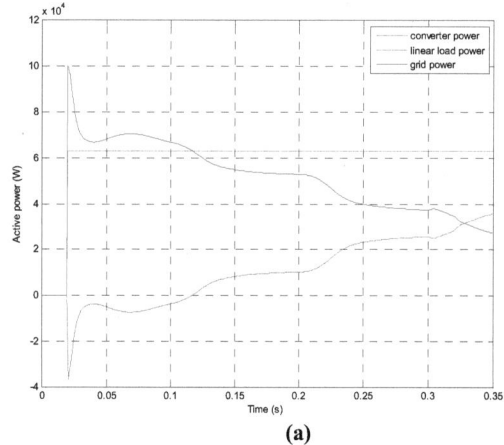

(a)

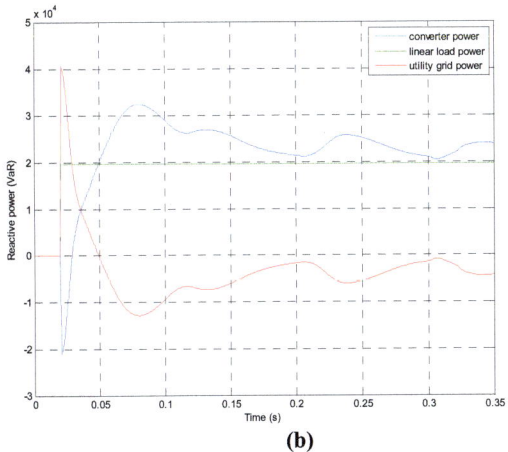

(b)

Figure 8.9: (a) Active power (b) Reactive power, variation of VSC, load and utility grid (Unity power factor, 49 Hz)

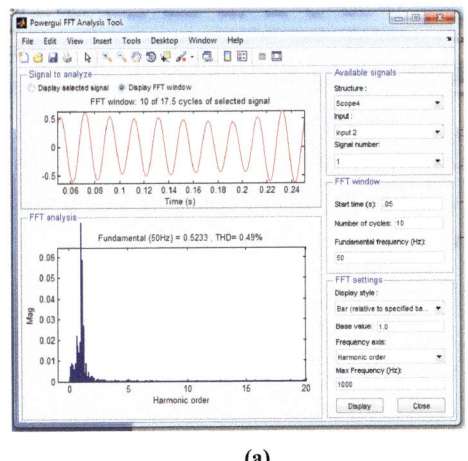

(a)

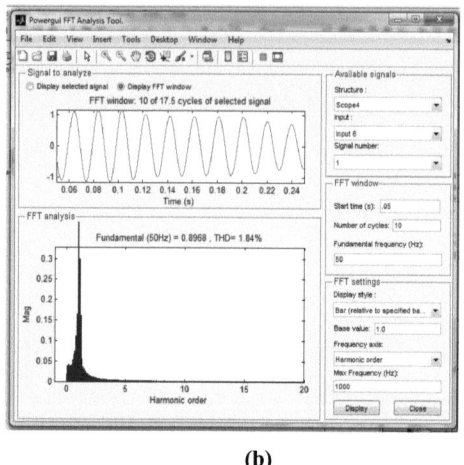

(b)

Figure 8.10: FFT analysis of (a) VSC current and (b) grid current (Unity power factor, 49 Hz)

- *at unity power factor, 440 V, 51 Hz*

Figure 8.11 (a) shows the real output power response whereas; Figure 8.11 (b) shows reactive output power response of VSC, load and electric grid. Figure 8.12 (a) and Figure 8.12 (b) gives the THD response of VSC current and grid injected current at unity power factor, respectively. However, this response has been obtained at 49 Hz frequency. The THD values are found to be 6.52% and 1.84%, respectively. This it has is concluded that change in frequency has no effect on the THD of grid current whereas, THD of VSC current has been increased from 0.49% to 6.52%.

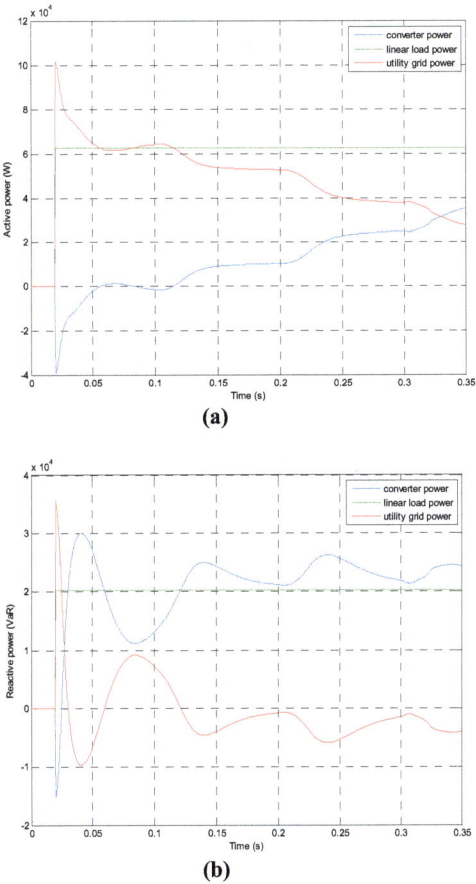

Figure 8.11: (a) Active power (b) Reactive power, variation of VSC, load and utility grid (Unity power factor, 51 Hz)

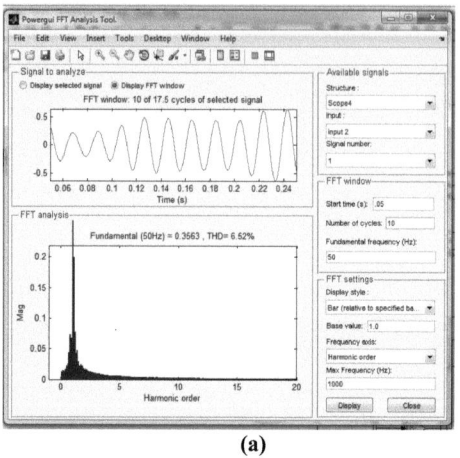

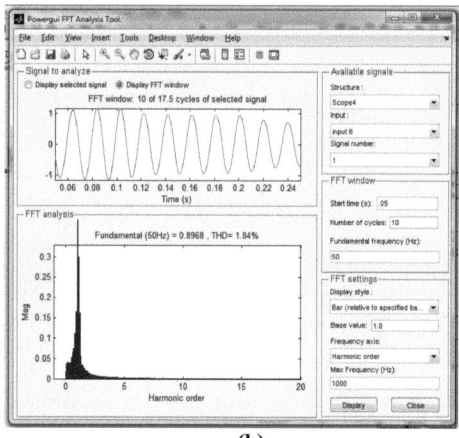

Figure 8.12: FFT analysis of (a) VSC current and (b) grid current (Unity power factor, 51 Hz)

8.3 Conclusion

In this chapter, the simulation of the proposed system is performed under two different cases. In first case, the effect of changing power factor on active power, reactive power and THD values is observed. It is observed that the THD of grid current increases with increase in the phase angle of grid current VSC voltage. It affects the active power flow among VSC, load and utility grid. In the second case, the effect of changing frequency on active power, reactive power and THD values is noticed. It is observed that the THD of VSC current increases whereas, the THD of grid current remains constant.

CHAPTER IX: MODEL ANALYSIS WITH INCREMENTAL CONDUCTANCE MPPT TECHNIQUE

9.1 Introduction

The principle and need of MPPT techniques has been presented in chapter 4. Out of all available MPPT techniques, two basic MPPT techniques have been discussed in chapter 4. In order to evaluate the performance of the proposed system, the incremental conductance MPPT technique has been implemented with two-stage solar PV grid connected system. In this case, the PCS consists of DC-DC boost converter as first stage and DC-AC IGBT converter as second stage. MPPT is implemented in the control scheme of the boost converter. In the second stage, the DC voltage and current values are controlled with the help of PI controllers. These controllers are of type voltage and current controlled type, respectively. The control scheme of the IGBT provides the gate pulse to it through the PWM generator.

Figure 9.1 shows the simulink diagram of incremental conductance MPPT technique in which PV current and PV voltage are the controlled variables. The initial duty cycle is assumed as unity for the DC-DC boost converter. The saturation block limits the signal range within the specified limits, Figure 9.2. Pulse generator (5 kHz) sends the pulse to the gate of the boost converter. Sine Pulse Width Modulation (SPWM) method is used for generation of the pulse for converter.

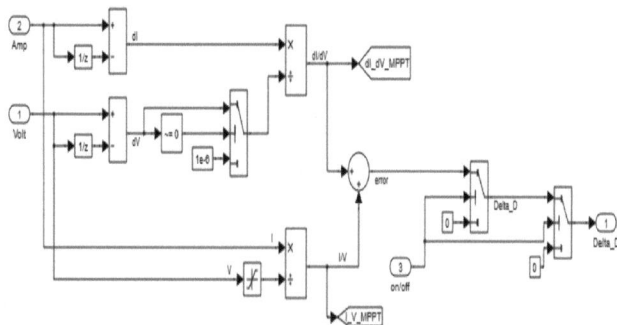

Figure 9.1: Simulink diagram of incremental conductance MPPT technique

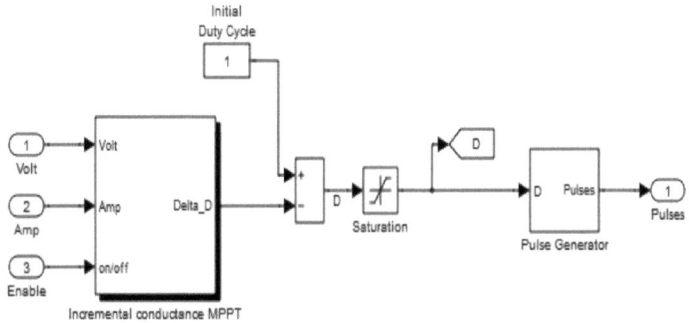

Figure 9.2: Simulink diagram of boost converter control

9.2 Voltage and Current Controllers

As depicted in Figure 9.3, the actual value of DC voltage to DC-AC converter is compared with the reference value of the MPPT control. This voltage is compared in the PI based voltage controller. The proportional error is passed through the PI controller, which generates the reference value of the direct-axis signal current. Figure 9.4 shows the PI based current controller in which the current signal is converted into the voltage signal. In order that the system is evaluated at unity power factor, quadrature component of the current is kept zero.

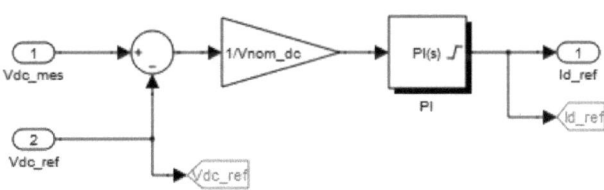

Figure 9.3: Simulink diagram of DC voltage PI controller

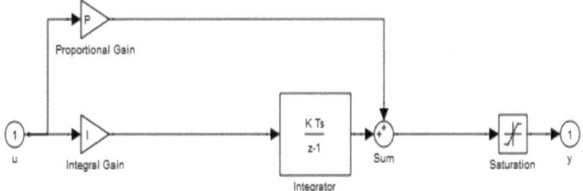

Figure 9.4: Simulink diagram of DC current PI controller

9.3 Solar PV Computation Model

According to the conducted literature survey, there are two types of solar PV conversion systems: single-stage and double-stage. As shown in Figure 9.5, the power conditioning system which is double-stage consists of a solar PV array, DC-DC boost converter with its MPPT control, and IGBT based DC-AC converter with PI control. The grid voltage and current from bus number 4 has been taken as the variables and fed to controller of DC-AC converter. In addition, the actual DC voltage to VSC is also to be controlled and maintained to track the reference MPPT voltage. The proposed system is tested at linear RLC load. In this chapter, IC based MPPT technique has been implemented. Figure 9.6 demonstrates the simulink tags for the measurement of voltage, current and power values. These values have been measured at VSC, load and utility grid side. Table 9.1 lists the specifications of each sub-system of the proposed simulink system. Table 9.2 lists the 100 kW PV module data-sheet specifications of manufacturer [19] (Sun Power SPR-305-WHT).

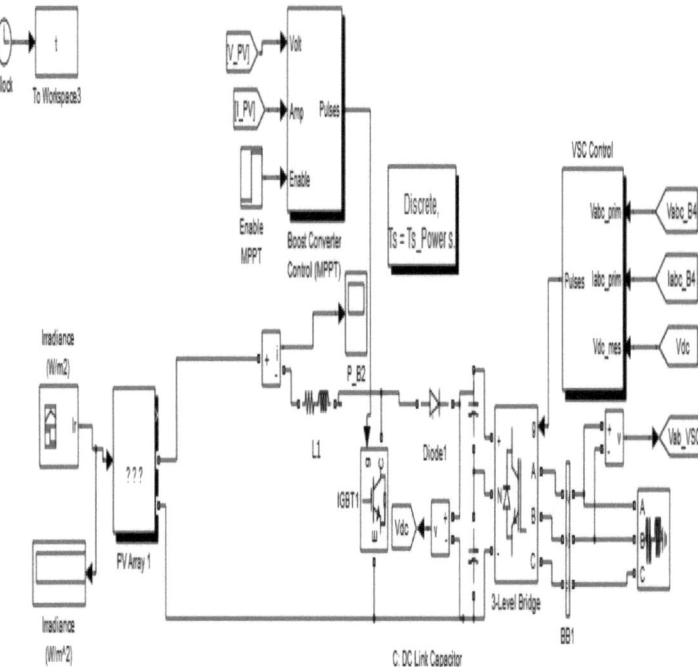

Figure 9.5: Simulink diagram of solar PV with connected load

Table 9.1: Parameters of PV grid connected system

System name	Rating values
Nominal DC voltage	200 V
Nominal power and frequency	200 kW, 50 Hz
DC voltage regulator gains (K_p, K_i)	9.5, 950
Current regulator gains (K_p, K_i)	0.55, 45
Sample time	50×10^{-6} s
LC filter (L, C)	1500 µH, 30 µF
3-φ transformer nominal power and frequency	200 kVA, 50 Hz
Load (V_n, P, Q_l)	440 V, 82 kW, 22 kVAR

Table 9.2: Specifications adopted for single PV array (Sun Power SPR-305-WHT) [19]

System name	Rating values
No. of cells per module	96
No. of series connected modules per string	5
No. of parallel strings	66
Module specifications under STC $[V_{oc}, I_{sc}, V_{mp}, I_{mp}]$	[64.2 V, 5.96 A, 54.7 V, 5.58 A]
Model parameters for one module $[R_s, R_p, I_{sat}, I_{ph}, Q_d]$	[0.038 Ω, 993.5 Ω, 3.1949×10^{-8} A, 5.9602 A, 1.3]
Maximum power P_{mp} per array	$66 \times 5 \times 54.7 \times 5.58$ = 100.7 kW per array

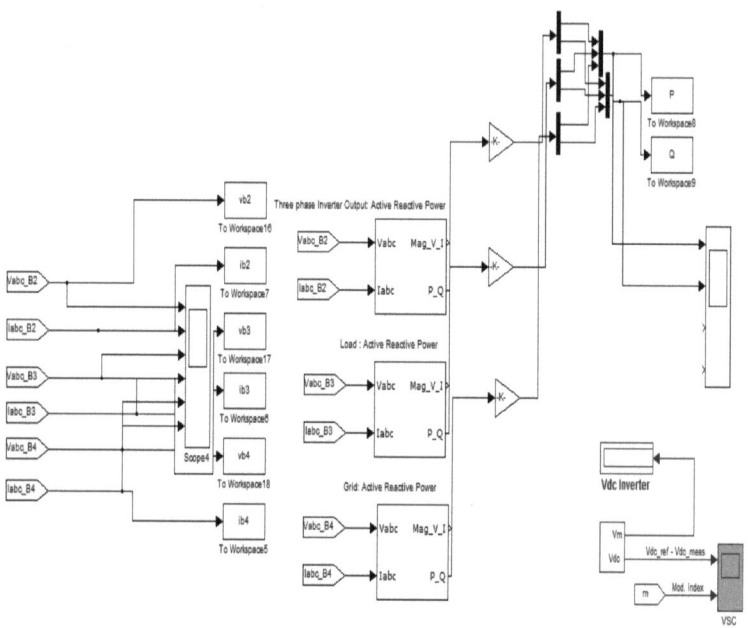

Figure 9.6: Different signal measurement blocks

9.4 Simulations Results and Discussion

In order to show the effectiveness of the proposed system, it has been tested at linear *RLC* load, where the total simulation time period is $t = 0.3$ s. The real and reactive power exchange has been demonstrated among PV-VSC, load and utility grid. The tracking behavior of actual DC link voltage around reference IC-MPPT voltage has been presented. Finally, power quality evaluation study has been carried by measuring the THD at PV-VSC, load and utility grid side.

Figure 9.7 (a)-(f) depicts the voltage and current waveform of VSC, connected load and utility grid. It has been observed from Figure 9.7 (b) that current starts flow from VSC at $t = 0.05$ s. From Figure 9.7 (d), it is found that the sine wave current flows through the connected load. Small harmonics are observed in the grid current waveform from Figure 9.7 (f). These harmonics are due to the current injection by the semiconductor devices like wise IGBT based VSC.

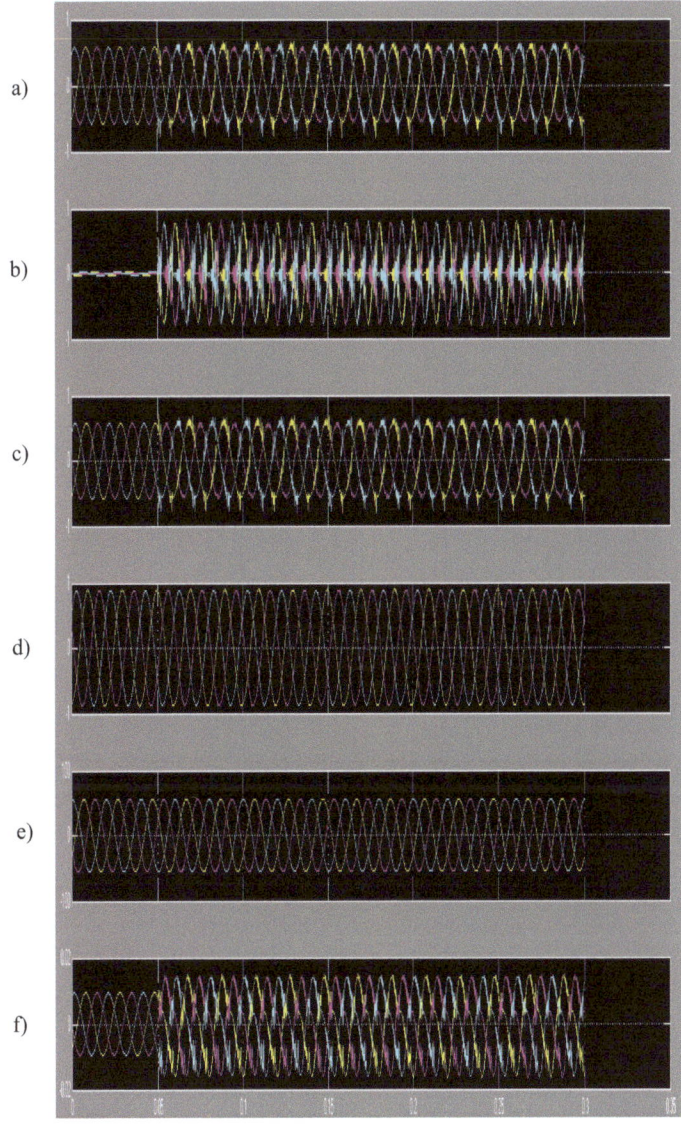

Figure 9.7: Waveform of a) VSC Voltage b) VSC Current c) Load Voltage d) Load Current e) Utility grid Voltage f) Utility grid Current

Figure 9.8 depicts the tracking of actual DC link voltage around the reference voltage of 200 V. It is possible to observe that actual behavior of DC link voltage to VSC is having more magnitude than the reference voltage. Small transients are started at $t = 0.07$ s, which can be controlled by the further tuning of PI based voltage and current controllers. Figure 9.9 shows that the modulation index remains unity during the simulation period.

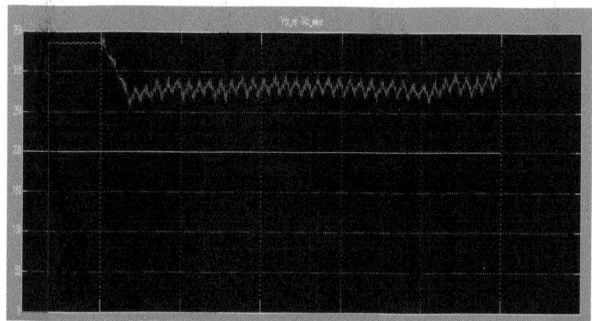

Figure 9.8: Change in actual DC link voltage with MPPT reference voltage

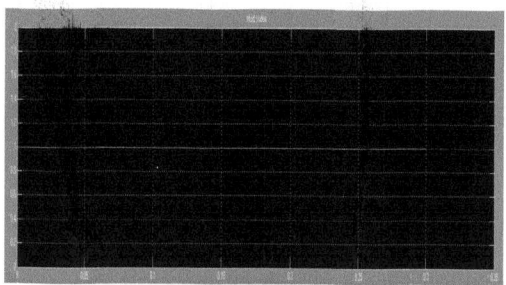

Figure 9.9: Modulation index

Figure 9.10 (a) depicts the active power response, whereas Figure 9.10 (b) depicts the reactive power response of PV array through VSC, connected load and utility grid.

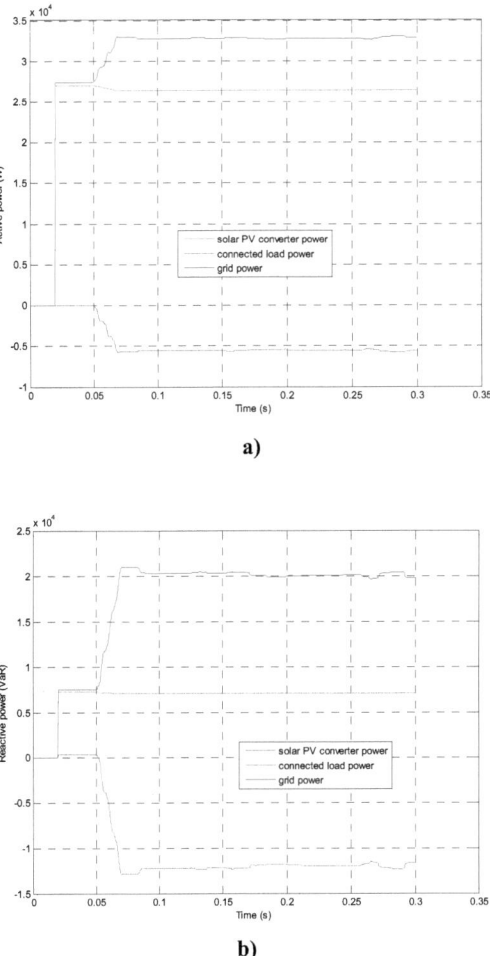

Figure 9.10: Waveform of a) Real power and b) Reactive power, with IC MPPT technique

In order to demonstrate the power quality evaluation of the proposed system, the THD measurement and its analysis is carried out with the FFT method. Figure 9.11 (a) and Figure 9.11 (b) shows the THD analysis of load current and grid current, respectively. The THD values of load current and grid current are found to be 0.93% and 28.94%, respectively. Table 9.3 shows the complete THD analysis from where it is found that THD of converter current is 85.77%. It is due to the fact that PV current flows through the VSC, which is a non-linear device.

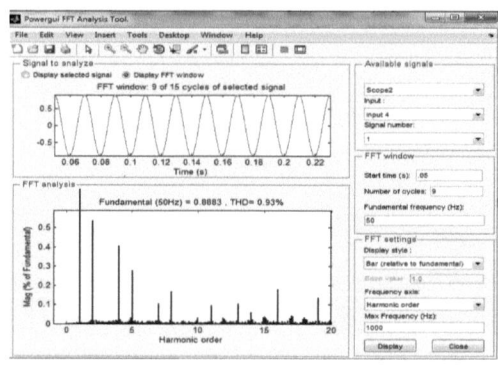

a)

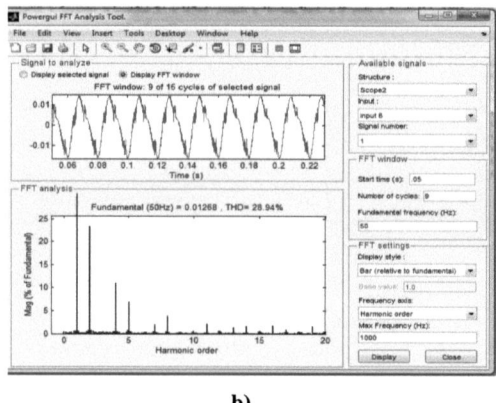

b)

Figure 9.11: THD analysis of a) load current and b) grid current

Table 9.3: Total harmonic distortion analysis using IC-MPPT

Parameter	THD (%)
DC-AC converter voltage	6.60
DC-AC converter current	85.77
linear-load voltage	6.60
linear-load current	0.93
Utility-grid voltage	0.22
Utility-grid current	28.94

9.5 Conclusion

This chapter has presented the analysis of two-stage solar PV grid connected system which is evaluated at linear *RLC* load. In DC-DC boost converter, the IC-MPPT technique which is capable to operate even under changing environmental conditions is implemented. Real and reactive power exchange is exchanged among VSC, load and utility grid. Voltage and current waveforms are presented. In order to evaluate the level of power quality, the THD analysis is carried out using FFT. It has been found that although level of harmonics generation from VSC is high, the control system is designed that level of harmonics is reduced for grid injected current.

CHAPTER X: MODEL ANALYSIS WITH PERTURB & OBSERVE MPPT TECHNIQUE

10.1 Introduction

In this chapter, the proposed system explained chapter 9 has been tested at P&O-MPPT technique. In the previous chapter, the solar PV grid connected system has been tested at IC-MPPT technique. The PI type voltage and current controllers have been implemented which are used for the VSC control. In the first stage, DC-DC boost converter is used whereas DC-AC converter is used as the second stage. The actual DC link voltage to VSC has also been studied whose behavior varies around the reference MPPT voltage. Finally, the THD analysis has been carried out using FFT analysis.

10.2 Proposed System Configuration

Figure 10.1 shows the simulink diagram of complete system. In the system tested in chapter 9, the IC based MPPT technique was used for extracting the power at MPP. This technique was found to be suitable under changing environmental conditions. Therefore, in order to validate the proposed system the system is tested at P&O based MPPT technique. The P&O based MPPT technique has been designed which is already explained in chapter 4. The VSC control has been accomplished by PI voltage and current controllers. The parameters chosen have been listed in Table 9.1 and Table 9.2 of chapter 9. A boost converter has been used which increases the voltage level. *LC* filter is used after the VSC. This filter is used and presents the injection of harmonics into the utility grid side.

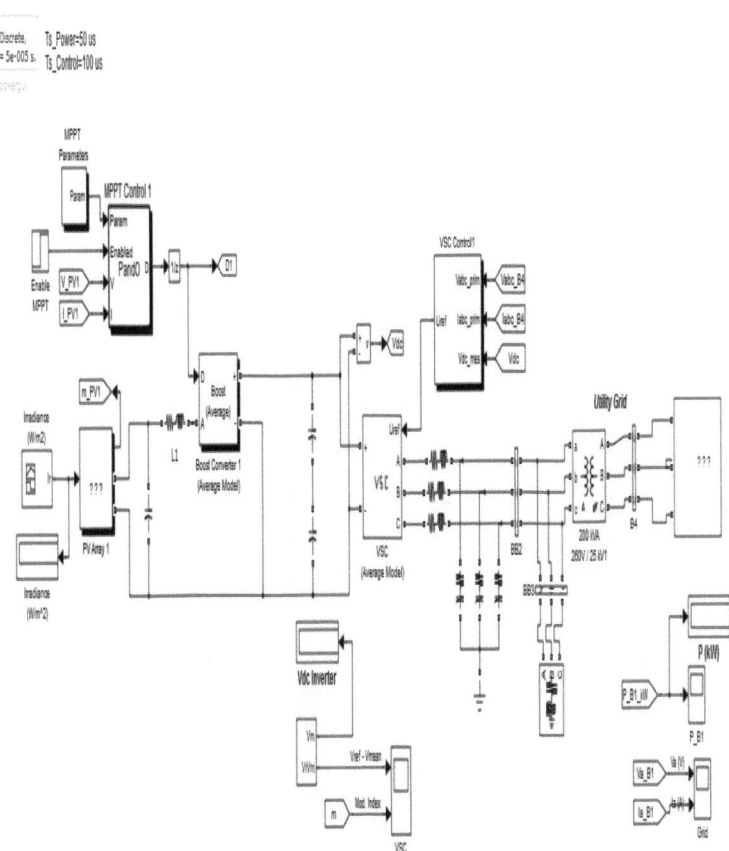

Figure 10.1: Simulink diagram of photovoltaic grid connected system

10.3 Simulation Results and Case Studies

In order to validate the results obtained in the previous chapter, this section presents the simulated results which have been obtained with P&O based MPPT technique. Figure 10.2 (a) depicts the waveform of VSC voltage which is sinusoidal and amplitude remains constant. Figure 10.2 (b) shows the waveform of VSC current whose value starts sine shape at 0.05 s. Figure 10.2 (c) depicts the waveform of load voltage whereas, Figure 10.2 (d) depicts the waveform of load current. Figure 10.2 (e) depicts the waveform of utility grid voltage and Figure 10.2 (f) depicts the waveform of utility grid current.

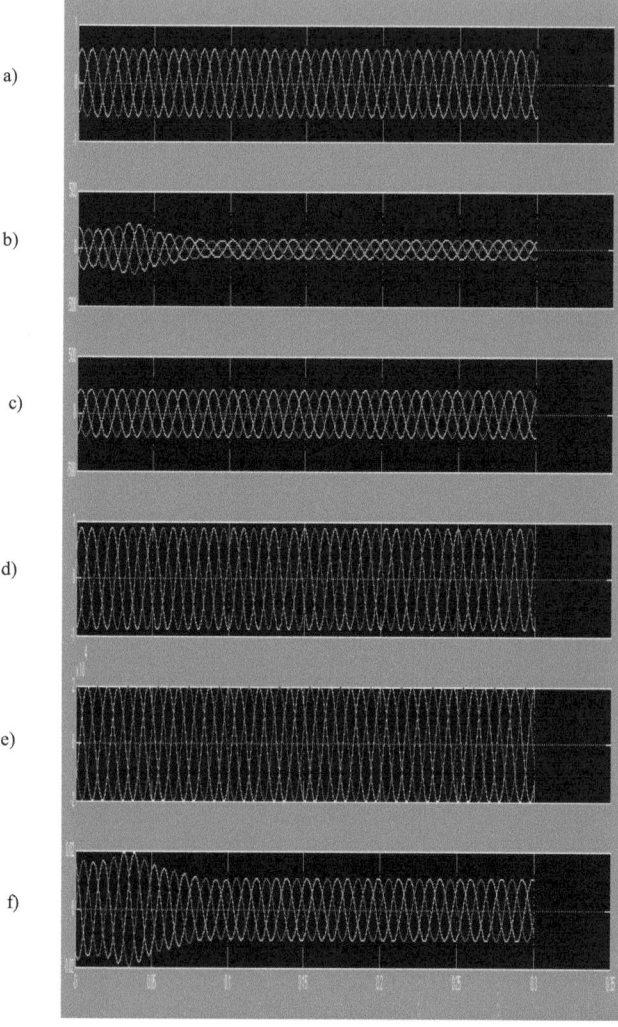

Figure 10.2: Waveform of a) VSC Voltage b) VSC Current c) Load Voltage d) Load Current e) Utility grid Voltage f) Utility grid Current

Figure 10.3 (a) depicts the waveform of active power for solar PV through VSC, load and utility grid. Figure 10.3 (b) depicts the waveform of reactive power from where it is clear that the VSC power and utility grid power are negative. Therefore, both are absorbing instead of generating the reactive power. The magnitude of reactive power depends upon the product of voltages of both utility grid and VSC, whereas the magnitude of real power depends upon the product of voltage and phase angle between grid voltage and VSC current.

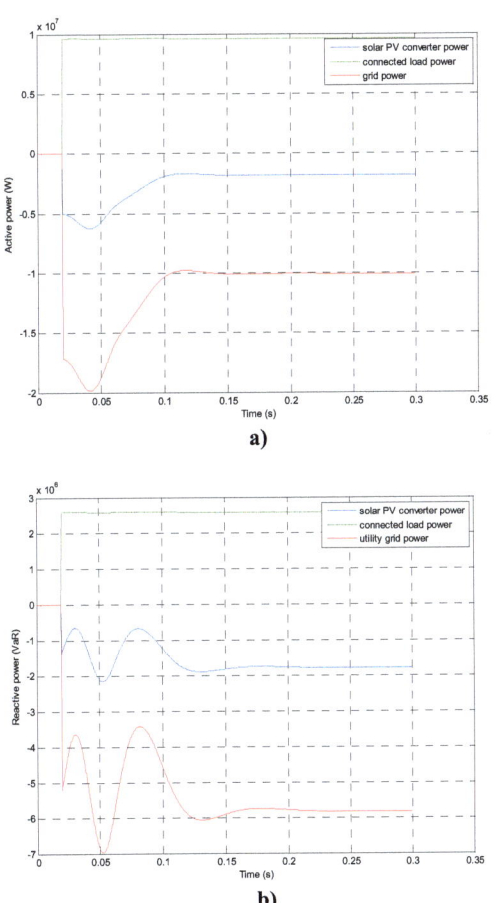

Figure 10.3: Waveform of a) Real power and b) Reactive power, with P&O MPPT technique

Figure 10.4 depicts the behavior of actual DC link voltage around the reference MPPT voltage (200 V). It is possible to observe that after 0.05 s, the actual DC link voltage doesn't track the reference voltage. This shows that this P&O-MPPT technique is not able to track the reference voltage under wide change of environmental conditions. Figure 10.5 shows the variation of modulation index with time.

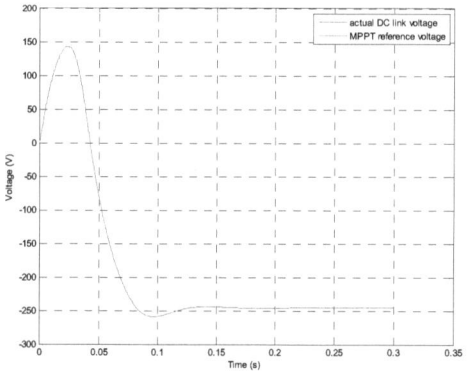

Figure 10.4: Change in actual DC link voltage with MPPT reference voltage

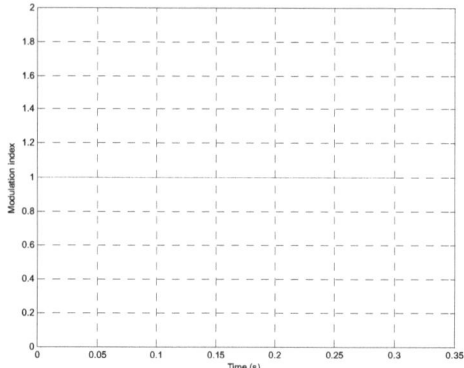

Figure 10.5: Modulation index

In order to demonstrate the power quality evaluation and validate the proposed system, the THD measurement and its analysis is carried out with the FFT method. Figure 10.6 (a) and Figure 10.6 (b) shows the THD analysis of load current and grid current, respectively. The THD values of load current and grid current are found to be 0.17% and 5.26%, respectively. Table 10.1 shows the complete THD analysis from where it is found that THD of converter current is 4.53%. It has been found to be lesser as compared to the value obtained with IC-MPPT technique.

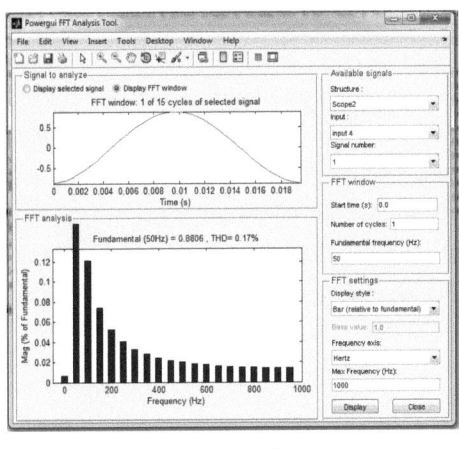

a)

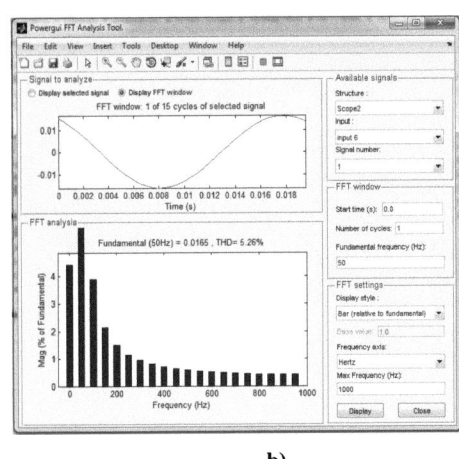

b)

Figure 10.6: THD analysis of a) load current and b) grid current

Table 10.1: Total harmonic distortion analysis using P&O MPPT

Parameter	THD (%)
DC-AC converter voltage	0.17
DC-AC converter current	4.53
linear-load voltage	0.17
linear-load current	0.17
Utility-grid voltage	0.01
Utility-grid current	5.26

10.4 Conclusion

This chapter has discussed the system performance at P&O based MPPT technique for solar PV two-stage grid connected system. It has been found that harmonic level is reduced for converter current and grid injected current. However, this MPPT is not able to track the reference MPPT voltage accurately. This validates that this MPPT is not able to operate under wide range of changing environmental conditions. The behavior of voltage and current levels of VSC, load connected and utility grid has also been discussed. Active and reactive power exchange among VSC, load connected and utility grid has also been highlighted.

References

[1] International Energy outlook. Chapter 5: Electricity. (May 2016). *US Energy Information Administration*. [Online] Available: http://www.eia.gov/forecasts/ieo/electricity.cfm

[2] Electricity Sector in India. (July 2016). [Online]. Available: https://en.wikipedia.org/wiki/Electricity_sector_in_India

[3] Power Sector in India-Solar Renewable and Wind Energy Sectors. *India Brand Equity Foundation*. (July 2016). [Online]. Available: http://www.ibef.org/industry/power-sector-india.aspx

[4] Renewables Global Status Report. REN21 *Renewable Energy Policy Network for the 21st Century* 2016.

[5] Ministry of New & Renewable Energy. (2016, June). Physical progress achievements. Government of India.

[6] R. A. Messenger, and J. Ventre, *Photovoltaic Systems Engineering*. New York: CRC Press, 2004.

[7] H. S. Rauschenbach, *Solar Cell Array Design Handbook-The Principles and Technology of Photovoltaic Energy Conversion*. New York: Van Nostrand Reinhold, 1980.

[8] J. A. Gow, and C. D. Manning, "Development of a model for photovoltaic arrays suitable for use in simulation studies of solar energy conversion systems", In: Proc. of 6th IEEE International Conference on Power Electronics and Variable Speed Drives, pp. 69-74, September 1996.

[9] J. A. Gow, and C. D. Manning, "Development of a photovoltaic array model for use in power electronics simulation studies", In: Proc. of IEE Electric Power Applications, vol. 146, no. 2, pp. 193-200, March 1999.

[10] F. Lasnier, and T. G. Ang, *Photovoltaic Engineering Handbook*. Bristol, UK: Adam Hilger Publishing, 1990.

[11] K. Khouzam, C. Ly, C. K. Koh, and P. Y. Ng, "Simulation and real-time modelling of space photovoltaic systems", In: Proc. of 24th IEEE World Conference on Photovoltaic Energy Conversion, Waikoloa, HI, vol. 2, pp. 2038-2041, December 1994.

[12] C. S. Solanki, *Solar Photovoltaics-Fundamentals, Technologies and Applications*. 2nd ed. New Delhi: PHI Learning, July 2011.

[13] M. Buresch, *Photovoltaic Energy Systems Design and Installation*. New York: McGraw-Hill, 1983.

[14] I. H. Altas, and A. M. Sharaf, "A PV array simulation model for Matlab-Simulink GUI environment", In: Proc. of IEEE International Conference on Clean Electrical Power (ICCEP), Capri, pp. 341-345, May 2007.

[15] M. A. Green, and S. R. Wenham, "Novel parallel multi-junction solar cell", *Applied Physics Letters,* vol. 65, pp. 2907, 1994.

[16] K. Zweibel, *Harnessing Solar Power*, Plenum Press, New York, 1990.

[17] J. Yang, A. Banerjee, T. Glatfelter, S. Sugiyama, and S. Guha, Conference record of the 26th IEEE Photovoltaics specialists Conference, pp. 563-568, 1997.

[18] V. A. Chaudhari, "Automatic peak power tracker for solar PV modules using dSPACE software, Dissertation of Master of technology in Energy, NIT Bhopal, 462 007.

[19] Manufacturer Datasheet Specifications, "PV model Sunpower SPR-305-WHT".

[20] W. Xiao, and W. G. Dunford, "A modified adaptive hill climbing MPPT method for photovoltaic power systems," In: Proc. of 35th International Conference on Annualized IEEE Power Electronics, pp. 1957–1963, 2004.

[21] A. A. Amoudi, and L. Zhang, "Optimal control of a grid-connected PV system for maximum power point tracking and unity power factor," In: Proc. of 7th International Conference on Power Electron. Variable Speed Drives, 1998, pp. 80–85.

[22] C. C. Hua, and J.R. Lin, "Fully digital control of distributed photovoltaic power systems," In: Proc. of IEEE International Conference on Industrial Electronics, pp. 1-6, 2001.

[23] G.M.S. Azevedo, M.C. Cavalcanti, K.C. Oliveira, F.A.S. Neves, and Z.D. Lins, "Comparative evaluation of Maximum power point tracking methods for photovoltaic systems" in *Journal of Solar energy engineering*, ASME, pp. 031006-1-- 031006-8, 2009.

[24] T. Esram, and P.L. Chapman, "Comparison of photovoltaic array maximum power point tracking techniques", *IEEE transactions on energy conversion,* vol. 22, no. 2, pp. 439-449, June 2007.

[25] G. W. Hart, H. M. Branz, and C. H. Cox, "Experimental tests of open loop maximum-power-point tracking techniques," *Solar Cells*, vol. 13, pp. 185–195, 1984.

[26] K. Kobayashi, H. Matsuo, and Y. Sekine, "A novel optimum operating point tracker of the solar cell power supply system," In: Proc. of 35th Annu. International Conference on IEEE Power Electronics, pp. 2147-2151, 2004.

[27] S. Yuvarajan, and S. Xu, "Photo-voltaic power converter with a simple maximum-power-point-tracker," In: Proc. of 2003 International Conference on Symposium on Circuits Systems, pp. III-399–III-402, 2003.

[28] A. Jain, "PI and fuzzy controller based DVR to mitigate power quality and reduce the harmonics distortion of sensitive load", ME Dissertation, Thapar University, July 2013.

[29] W. Raithmayr, P. Daehler, M. Eichler, G. Lochner, E. John, and K. Chan, "Customer Reliability Improvement with a DVR or a DUPS", Power World 98, pp. 1-10, 1998.

[30] B. M. Das, and G.K. Dubey, "Performance Study of UPQC-Q for Load Compensation and Voltage Sag Mitigation", Proc. of IEEE 28th Annual Conference of the Industrial Electronics Society (IECON 02), vol. 1, pp. 698-703, 2002.

[31] Y. Chen, and K. M. Smedley, "A cost-effective single-stage inverter with maximum power point tracking", *IEEE Trans. on Power Electronics*, vol. 19, no. 5, pp.1289-94, Sep. 2004.

[32] H. Moin, "Investigation to improve the control and operation of a three-phase PV grid-tie inverter", PhD Dissertation, Dublin Institute of Technology, pp. 1-248, 2011.

[33] I. Colak, E. Kabalci, and G. Bal, "Parallel DC-AC conversion system based on separate solar farms with MPPT control", In: Proc. of 8th IEEE International Conference on Power Electronics and ECCE Asia, Jeju, pp. 1469-1475, June, 2011.

About the Author

Dr. Akhil Gupta received B.E. (Electrical Engineering) from Giani Zail Singh-Punjab Technical University Campus, Bathinda (affiliated with Inder Kumar Gujral PTU, Jalandhar, India) in 1999, and M.Tech. in Electrical Engineering from Kay Jay group of Institutes, Patiala (Institute of Advanced Studies In Education, Rajasthan) in 2005. He completed his Ph.D. from the Department of Electrical Engineering, National Institute of Technology,Kurukshetra, Haryana, India, in 2016. He is now Associate Professor in the Department of Electrical Engineering at Chandigarh University, Gharuan, Mohali, Punjab, India. He has around 17 years of experience in academics/industry and has published more than 40 publications in reputed International Journals/Conferences. His areas of research are integration of solar photovoltaic energy systems into electrical power systems, power quality and control.